SpringerBriefs in Applied Sciences and Technology

Nanoscience and Nanotechnology

Series editor

Hilmi Volkan Demir, Nanyang Technological University, Singapore, Singapore

Nanoscience and nanotechnology offer means to assemble and study superstructures, composed of nanocomponents such as nanocrystals and biomolecules, exhibiting interesting unique properties. Also, nanoscience and nanotechnology enable ways to make and explore design-based artificial structures that do not exist in nature such as metamaterials and metasurfaces. Furthermore, nanoscience and nanotechnology allow us to make and understand tightly confined quasi-zero-dimensional to two-dimensional quantum structures such as nanoplatelets and graphene with unique electronic structures. For example, today by using a biomolecular linker, one can assemble crystalline nanoparticles and nanowires into complex surfaces or composite structures with new electronic and optical properties. The unique properties of these superstructures result from the chemical composition and physical arrangement of such nanocomponents (e.g., semiconductor nanocrystals, metal nanoparticles, and biomolecules). Interactions between these elements (donor and acceptor) may further enhance such properties of the resulting hybrid superstructures. One of the important mechanisms is excitonics (enabled through energy transfer of exciton-exciton coupling) and another one is plasmonics (enabled by plasmon-exciton coupling). Also, in such nanoengineered structures, the light-material interactions at the nanoscale can be modified and enhanced, giving rise to nanophotonic effects.

These emerging topics of energy transfer, plasmonics, metastructuring and the like have now reached a level of wide-scale use and popularity that they are no longer the topics of a specialist, but now span the interests of all "end-users" of the new findings in these topics including those parties in biology, medicine, materials science and engineerings. Many technical books and reports have been published on individual topics in the specialized fields, and the existing literature have been typically written in a specialized manner for those in the field of interest (e.g., for only the physicists, only the chemists, etc.). However, currently there is no brief series available, which covers these topics in a way uniting all fields of interest including physics, chemistry, material science, biology, medicine, engineering, and the others.

The proposed new series in "Nanoscience and Nanotechnology" uniquely supports this cross-sectional platform spanning all of these fields. The proposed briefs series is intended to target a diverse readership and to serve as an important reference for both the specialized and general audience. This is not possible to achieve under the series of an engineering field (for example, electrical engineering) or under the series of a technical field (for example, physics and applied physics), which would have been very intimidating for biologists, medical doctors, materials scientists, etc.

The Briefs in NANOSCIENCE AND NANOTECHNOLOGY thus offers a great potential by itself, which will be interesting both for the specialists and the non-specialists.

More information about this series at http://www.springer.com/series/11713

Alexander Govorov
Pedro Ludwig Hernández Martínez
Hilmi Volkan Demir

Understanding and Modeling Förster-type Resonance Energy Transfer (FRET)

Introduction to FRET, Vol. 1

 Springer

Alexander Govorov
Department of Physics and Astronomy
Ohio University
Athens, OH
USA

Pedro Ludwig Hernández Martínez
School of Physical and Mathematical
  Sciences, LUMINOUS! Centre of
  Excellence for Semiconductor Lighting
  and Displays, TPI—The Institute of
  Photonics
Nanyang Technological University
Singapore
Singapore

Hilmi Volkan Demir
Department of Electrical and Electronics
  Engineering, Department of Physics, and
  UNAM—National Nanotechnology
  Research Center and Institute of Materials
  Science and Nanotechnology
Bilkent University
Ankara
Turkey

and

School of Electrical and Electronic
  Engineering, School of Physical and
  Mathematical Sciences, LUMINOUS!
  Center of Excellence for Semiconductor
  Lighting and Displays, TPI—The Insitute
  of Photonics
Nanyang Technological University
Singapore
Singapore

ISSN 2191-530X   ISSN 2191-5318   (electronic)
SpringerBriefs in Applied Sciences and Technology
ISSN 2196-1670   ISSN 2196-1689   (electronic)
Nanoscience and Nanotechnology
ISBN 978-981-287-377-4   ISBN 978-981-287-378-1   (eBook)
DOI 10.1007/978-981-287-378-1

Library of Congress Control Number: 2016943801

Printed on acid-free paper

This Springer imprint is published by Springer Nature
The registered company is Springer Science+Business Media Singapore Pte Ltd.

Contents

Chapter 1
Short History of Energy Transfer Theory Before Förster, At The Time of Förster, and After Förster

This chapter gives a brief introduction to the historical development of energy transfer. In this chapter, we are not dealing with a detailed review of applications, nor a review of modern theoretical development; instead, we outline some of the ideas, experiments and theories that formed the scientific background that culminated in the Förster resonance energy transfer (FRET) theory. A more detailed description of the historical events regarding to FRET can be found in a review [1] and references within.

FRET is a physical process where the excited state energy of the "donor" can be transferred to the "acceptor" in the ground state. This can take place whenever the donor and the acceptor are close enough, usually less than 10 nm at room temperature, and under certain other conditions.

FRET is one of the major experimental methods for discovering whether two interacting particles are in close proximity, or for determining the distance between two specific locations in a complex micro/nano structure. It has become a major experimental technique in the field of single particles, since the efficiency of energy transfer is measured via fluorescence tools. The energy transfer is typically detected relatively easily and it is used to show interactions between two particles. It is a powerful technique since FRET measures dynamics on a spatial scale that is unique (in nanometers range). It is fairly simple and can be studied in most of the laboratories. FRET has been used since early 1950s, however, its use has been exploded in the last decades mainly because of instrumentation improvements and innovations and the great number of commercially available synthetic fluorophores, which can be featured with particular chemical groups for specific purposes.

In the following section we briefly discuss the historical events that led to the understanding and modeling of FRET. First, the concept of electromagnetism and quantum mechanics as well as the concept of energy transfer previous to Förster are discussed. Next, we shortly review the experiments and concept of energy transfer

© The Author(s) 2016
A. Govorov et al., *Understanding and Modeling Förster-type Resonance
Energy Transfer (FRET)*, Nanoscience and Nanotechnology,
DOI 10.1007/978-981-287-378-1_1

during the Förster time. Finally, we summarize the concept of energy transfer after Förster.

1.1   Brief Review of Scientific Achievements Before Förster Theory

The notion that electricity and magnetism are related were suspected before 1800 because of the formal similarities between static electricity and magnetism. Hans Christian Oerstead in 1820 was the first to demonstrate the interrelationship of electricity and magnetism. He reported that a magnet's needle held next to a current-carrying wire was deflected and oriented itself perpendicular to the line of current. This discovery, easy to reproduce, was repeated by André-Marie Ampére and others (e.g., Jean-Baptiste Biot and Félix Savart). In 1822, Ampére published a theory of electromagnetic interactions involving currents. He found that current-carrying wires attract or repel each other depending on whether the currents is in the same direction or opposite.

Faraday who discovered that changing magnetic fields produce circulating electrical fields, known as "Faraday induction", introduced in 1821 an intuitive pictorial representation of lines of forces. Faraday pictured these lines of forces as the mechanism by which electrical and magnetic objects interact with themselves and with each other, where these lines of forces serve as "the carries" of forces through space. The first field theory was introduced by James Clerk Maxwell in 1864. His equations describe electromagnetic fields where the objects (electrical or magnetic) enter only through boundary conditions. Using the ideas and experiments of Faraday, Maxwell created a complete mathematical representation of Faraday's descriptions of electricity and magnetism. He understood Faraday's lines of force as a line passing through any point of space representing the direction of the force exerted giving the vector representation of the electromagnetic field. In addition to Faraday's ideas, Maxwell introduced the notion of displacement current, that is, the circulatory magnetic field caused by a time-varying electric field. Maxwell's equations describe all classical electrodynamic phenomena and settled the theoretical basis to predict electromagnetic radiation, which is the starting point for describing the classical theory of energy transfer.

Maxwell's equations predicted the identity of electromagnetism and light and the quantitative properties of light (interference, refrangibility and polarization as well as the speed of light). This was confirmed by the experiment of Heinrich Hertz with his famous Hertzian oscillating dipole. Hertz carried out the experiments in 1888 and in 1889 he published the theory to explain the electromagnetic fields surrounding this electric oscillator. These theory was derived from Maxwell's equations. Hertz experiments were performed by producing high frequency repetitive sparks in an air gap of a primary oscillating circuit. The electrodynamics disturbance was detected at a distance by a secondary circuit, resonance with the first, with similar air gaps. Sparks were observed in the secondary receiving circuit when it was resonant with the primary circuit. Hertz calculated a graphic field-line representation of the electromagnetic field of an oscillating dipole in the near field (shorter than a wavelength of the emitted radiation), in an intermediate zone, and in the far field (at distances farther than a wavelength) where the electromagnetic energy escapes as radiation. Hertz graphical and mathematical descriptions of the oscillating electric field emanating from his Hertzian dipole, in particular in the near field, played a critical role in understanding of the energy transfer theory.

The first recorded measurements of energy transfer over distances larger than collision radii were made by Cairo and Franck in 1922 [2–4]. Cairo observed emission from thallium in a mixture of mercury and thallium vapor, when the vapor mixture was excited at wavelength of 253.6 nm, which can only excite the mercury atoms. This fluorescence emission from thallium was named "sensitized fluorescence". The importance of resonance between energy levels of the sensitizer (the donor) and the sensitize (the acceptor) atoms was explicitly shown by further experiments, especially by the experiments of Beutler and Josephi [5, 6], who studied the sensitized fluorescence of sodium vapor in the presence of mercury vapor. The sensitized fluorescence increased in intensity when the energy differences are smaller between the states of the two participating atoms. Many of these experiments were interpreted in terms of collision theory. The number of collisions per time could be calculated from gas theory and the fraction of collision that was effective could be determined. If the rate of collisions is smaller than the calculated one, then only a certain percentage of the collisions are effective. If the rate of collisions is larger than the calculated, then there are interactions between the two collision partners that extend beyond their encounter radius. This discovery that energy transfer could take place over distances longer than the encounter radii showed that hard physical collisions were not required for atoms to interchange energy.

In 1928, Kallmann and London [7] proposed a quantum mechanical theory to explain the transfer of energy between atoms at longer distances compared to the collisional radii. This theory assumed *almost resonance* between the energy levels of the interacting atoms. They found that, provided that the corresponding transitions between the energy states of the two atoms were *dipole-allowed*, the effective cross-section q of the two interacting atoms increases as $\sigma^{-2/3}$, where σ is the difference between the excitation energies of the two interacting systems. As $\sigma \rightarrow 0$, the cross-section approaches a limiting value much larger than the collisional radii.

The Perrins (the father and the son) were the first to attempt a quantitative description of nonradiative energy transfer in solution between an excited molecule and a neighboring molecule in the ground state. The Perrins reasoned that the depolarization decrease that occurs in a fluorophore solution of at higher concentrations resulted from the transfer of excitation energy between molecules with different orientations before a photon was emitted. The Perrins' model involved a near-field energy interaction between the oscillating dipoles of two identical molecules, i.e., the oscillating dipoles are in resonance. Initially, J. Perrin (the father) developed a classical model to explain the depolarization decrease in a solution of a single chemical species of the fluorophore [8, 9]. He hypothesized that the transfer of the excitation energy could hop from one molecule to the other through interaction between oscillating dipoles of closely spaced molecules. J. Perrin modeled the participating molecules classically as Hertzian dipoles under the assumption that, if the molecules were separated by a sufficiently small distance, the energy could be transferred to the acceptor molecule nonradiatively. He named this process *"transfert d'activation"* (transfer of activation). He calculated that the distance for this process to take place is approximately $\lambda/(2\pi)$, where λ is the wavelength of the free electric field oscillating at the frequency of the atomic electric field, $\nu = c/\lambda$ (c is the speed of light). However, this value was 20-fold greater than the experimental results. Later, F. Perrin (J. Perrin's son) extended J. Perrin's theory of energy transfer by developing a quantum mechanical model [10, 11], similar to what had been suggested for the energy transfer between atoms in gases [7]. He estimated that the rate of transfer is proportional to $1/R^3$. This results in energy transfer at much longer distances than those found experimentally. Also later, F. Perrin considered collisions between the chromophores and the solvent molecules as well as Doppler effects. These reduced the distances to about 20–25 nm, which were still too long. Because of this discrepancy, the Perrin's theory of energy transfer lay dormant for about 20 years.

1.2 Förster Energy Transfer Theory

Förster's theory and his accompanying experimental work on the energy transfer are the most widely known, and most influential, of all energy transfer publications. His major papers are listed here [12–22]. T. Förster provided an accessible theory in a form that was practical for experimenters, commonly referred to as the Förster Resonance Energy Transfer (FRET) today. Because of this reason, FRET has been widely used in physics, engineering, chemistry, biology, and medicine.

T. Förster became interested in the energy transfer process because of the known effectiveness of photosynthesis. Experiments had shown that the capture and utilization of light energy by the plant's leaves was much more effective than was expected if it were required that photons exactly hit the reaction centers. He reasoned that an efficient transfer of energy between the chlorophyll molecules must be responsible for the eventual diffusion of the energy, which was absorbed over the whole surface of the leaf, into the relatively sparse reaction centers. Förster assumed that this diffusion is because of the energy being rapidly hopping (resonating) between molecules.

In his first paper on FRET [12], he correctly developed the basic theoretical background of FRET. First, he reviewed the mechanisms proposed by the Perrins. Then, he proceeded to take three critical, important steps that allowed him to derive a quantitative theory of nonradiative energy transfer [14]: (1) Förster took into account the broad spectral dispersion of the donor fluorescence and the acceptor absorption, i.e., the overlapping oscillation frequencies of the donor in the excited state and the acceptor molecules in the ground state. In his first paper [12], Förster treated this frequency overlap semi-classically and semi-quantitatively. However, shortly after [13, 14], he gave a full quantum mechanical treatment. (2) Förster was able to develop a quantitative theory of the rate of energy transfer from an excited donor molecule to a ground state acceptor molecule in terms of the *overlap integral*. The overlap integral represents the probability that the two molecular transition dipoles will have the same frequency. This was a major conceptual step because these spectroscopic transitions can be measured experimentally, independent of the FRET measurement, opening a way to quantitative interpretation of the experimental data. Förster also introduced helpful expressions for the orientation factor, κ^2, and included the effect of the index of refraction, which affects all electric interactions in condensed media at high optical frequencies. (3) Förster's model included quantitatively the $1/R^6$ distance dependence of the dipole-dipole interaction. He calculated the distance R_0 (known as the Förster radius) where the rate of the energy transfer was equal to the rate of fluorescence emission in terms of the overlap integral, the quantum yield of the acceptor, the lifetime of the donor in the absence of an acceptor and the effective index of refraction.

Förster's original theoretical description of energy transfer set the stage for all subsequent applications of FRET in many fields of research, and his theory is still used to interpret experimental results. His insight and great contribution were to provide the quantitatively correct and very practical description of the FRET process in terms of experimentally accessible parameters. By relating the rate of energy transfer to purely experimentally available parameters, he provided the general theoretical framework for all FRET applications.

It is worth mentioning that, before T. Förster, J.R. Oppenheimer reported the theory of FRET in 1941 at the American Physical Society meeting in a paper entitled "Internal Conversion in Photosynthesis" [23]. However, the full contribution of J.R. Oppenheimer and W. Arnold was published in 1950 [24]. In the 1941 abstract, Oppenheimer discussed that the high efficiency of the energy transfer from certain dyes to chlorophyll cannot happen due to light emission and re-absorption because the probability of this process is too small. However, the energy transfer can be enhanced if the chlorophyll molecules are much closer than the wavelength of chlorophyll fluorescence, that is, in the near field of a Hertzian dipole. In the 1950 publication, Arnold and Oppenheimer proposed a mechanism of the energy transfer from phycocyanin to chlorophyll in the blue-green algae. Here, they considered three ways for the energy transfer: (1) by direct collision, (2) by emission and re-absorption, and (3) by "internal conversion". They found that the probability for energy transfer by the first two mechanisms was too small. Therefore, they focused on the energy transfer in the near field zone of Hertzian dipole radiation, that is, "internal conversion". They calculated that the total energy transfer from phycocyanin to a randomly localized chlorophyll is proportional to $1/d^3$, which was the result of integrating $1/r^6$ from d to infinity.

1.3   Developments After Förster

Following the pioneering studies of the Perrins and Förster, there have been many
reports and reviews on FRET, both theoretical and experimental. A long literature
list is available in several recent reviews [25–33]. There have been several exten-
sions of the theory of energy transfer to other experimental conditions, by
T. Förster, D. Dexter, and others. Dexter [34] made a very important contribution
by generalizing and extending Förster's energy transfer model to include the donors
and acceptors with overlapping electron orbitals. This resulting energy transfer over
a very small distance is called "Dexter transfer" or transfer by "electron exchange".
The distance dependence of the Dexter transfer is very different from the Förster
transfer, and the rate of the Dexter process is only efficient for very short
donor-acceptor separation (<1 nm).

In Ref. [28], Medintz and Mattoussi provided an overview on FRET using
semiconductor quantum dots (QDs) and the application of QD-based FRET in
biology. They started by discussing some of the relevant conceptual elements of
FRET, the unique QD optical properties, and the advantages and limitations of
using QDs as exciton donors and/or acceptors. Then, they described representative
examples where QD-based FRET has been used for biological applications,
including the detection of hybridation using QD-nuclei acid conjugates, pH and ion
sensing, and antibody-based sensing. Overall, they provided a good understanding
of the most important parameters that govern FRET for the D-A pairs of QD-QD,
QD-dye, dye-QD, QD-Au-NP, and bioluminescent substrate-QD. Another review
by U.O.S. Seker and H.V. Demir presented a summary on FRET based systems and
applications using material binding peptides [29]. This focused on the selection
process, molecular binding characterization and utilization of peptides as molecular
linkers, molecular assembles and material synthesizers for FRET applications.

Rogach et al. [30] reviewed energy transfer using semiconductor nanocrystals
(NCs). This review semiconductors NC containing thin films, solution-based
complexes, and bioconjugates. Here, the energy transfer involving metal nanoan-
tennae and metal nanoparticles were discussed. In this review, it was concluded that
energy transfer involving semiconductor NCs coupled to metal nanoparticles or
dyes can be used for bio-imaging and sensing. And, the use of directional energy
transfer in semiconductor NCs can provide a new approach for hybrid photo-
voltaics. Agranovich et al. [31] presented a review on FRET in hybrid
organic-inorganic nanostructures. They reported several theoretical aspects of
energy transfer and discussed how hybrid organic-inorganic nanostructures can be
used for optoelectronic applications. A perspective on the recent understanding of
the excitonic dynamics in the organic-inorganic composites of semiconductor NCs
is given by Guzelturk and Demir [35]. In another review, Guzelturk et al. [32]
discussed the use of colloidal quantum dots and quantum wires in FRET for the
light generation and harvesting applications.

This chapter has covered mostly the work relevant to the understanding of energy transfer prior to Förster, leading up to the final, practical expression for FRET. In the following chapters we will discuss the process of energy transfer in detail.

References

1. M. Roberts, Clegg, Chap. 1, in *Reviews in Fluorescence 2006*, vol. 3, ed by C.D. Geddes, J.R. Lakowicz (Springer, Berlin, 2006)
2. G. Cairo, Über Entstehung wahrer Lichtabsorption un scheinbare Koppelung von Quantensprüngen. Z. Physik **10**, 185–199 (1922)
3. G. Cairo, J. Franck, Über Zerlegugen von Wasserstoffmolekülen durch angeregte Quecksilberatome. Z. Physik **11**, 161–166 (1922)
4. J. Franck, Einige aus der Theorie von Klein und Rosseland zu ziehende Folgerungen über Fluorescence, photochemische Prozesse und die Electronenemission glühender Körper. Z. Physik **9**, 259–266 (1922)
5. H. Beutler, B. Josephi, Resonanz by Stössen zweiter Art in der Fluoreszenz und Chemilumineszenz. Naturwuss **15**, 540 (1927)
6. H. Beutler, B. Josephi, Resonanz by Stössen zweiter Art in der Fluoreszenz und Chemilumineszenz. Z. Phys. **53**, 747 (1929)
7. H. Kallmann, F. London, Über quantenmechanische Energieübertragungen zwischen atomaren Systemen. Z. Physik. Chem **B2**, 207–243 (1928)
8. J. Perrin, *Fluorescence et radiochimie Conseil de Chemie*, Solvay, 2ièm, 1924 (Gauthier & Villar, Paris, 1925), pp. 322–398
9. J. Perrin, Fluorescence et induction moleculaire par resonance. C.R. Hebd. Seances Acad. Sci. **184**, 1097–1100 (1927)
10. F. Perrin, *Théorie quantique des transferts d'activation entre molécules de méme espèce. Cas des solutions fluorescentes*, Ann. Chim. Phys. (Paris) **17**, 283–314 (1932)
11. F. Perrin, Interaction entre atomes normal et activité. Transferts d'activation. Formation d'une molécule activitée. Ann. Institut Poincaré **3**, 279–318 (1933)
12. T. Förster, Energiewanderung und Fluoreszenz. Naturwissenschaften. **6**, 166–75 (1946)
13. T. Förster, Fluoreszenzversuche an Farbstoffmischungen. Angew. Chem. A **59**, 181–7 (1947)
14. T. Förster, Zwischenmolekulare Energiewanderung und Fluoreszenz. Ann. Phys. **2**, 55–75 (1948)
15. T. Förster, Expermentelle und theoretische Untersuchung des zwischengmolekularen Ubergangs von Elektronenanregungsenergie. A. Naturforsch. **4A**, 321–327 (1949)
16. T. Förster, Versuche zum zwischenmolekularen Ubergangs von Elektronenanregungsenergie. Z. Elektrochem. **53**, 93–100 (1949)
17. T. Förster, Fluoreszenz Organischer Verbindungen (Vandenhoeck & Ruprecht, Göttingen, 1951), 315p
18. T. Förster, Transfer mechanisms of electronic excitation. Discuss. Faraday Soc. **27**, 7–17 (1959)
19. T. Förster, Transfer mechanisms of electronic excitation energy. Radiat. Res. Suppl. **2**, 326–339 (1960)
20. T. Förster, *Delocalized excitation and excitation transfer*. Part III: action of light and organic molecules (Academic Press, New York, 1965), pp. 93–137
21. T. Förster (ed.), Delocalized excitation and excitation transfer, in *Modern quantum chemistry* ed. by O. Sunanoglu (Academic, New York, 1965), pp. 93–137
22. T. Förster, Intermolecular energy migration and fluorescence, in *Biological physics* ed. by E.V. Mielczarek, E. Greenbaum, R.S. Knox (American Institute of Physics, New York, 1993) pp. 148–160

23. J.R. Oppenheirmer, Internal conversion in photosynthsis. Phys. Rev. **60**, 158 (1941)
24. W. Arnold, J.R. Oppenheimer, Internal conversion in the photosynthetric mechanism of blue-green algae. J. Gen. Physiol. **33**, 423–435 (1950)
25. R.M. Clegg, Fluorescence resonance energy transfer, in *Fluorescence imaging. spectroscopy and microscopy*, vol. 137, ed. by X.F. Wang, B. Herman (Wiley, New York, 1996), pp. 179–252
26. W.B. Van Der Meer, G.I.I.I. Coker, S.-Y. Chen, *Resonance energy transfer: theory and data* (Wiley, New York, 1994)
27. P. Wu, L. Brand, Resonance energy transfer: methods and applications. Anal. Biochem. **218**, 1–13 (1994)
28. I.L. Meditz, H. Mattoussi, Quantum dot-based resonance energy transfer and its growing application in biology. Phys. Chem. Chem. Phys. **11**, 17–45 (2009)
29. U.O.S. Seker, H.V. Demir, Material binding peptides for nanotechnology. Molecules **16**, 1426–1451 (2011)
30. A.L. Rogach, T.A. Klar, J.M. Lupton, A. Meijerink, J. Feldmann, Energy transfer with semiconductor nanocrystals. J. Mater. Chem. **19**, 1208–1221 (2009)
31. V.M. Agranovich, Y.N. Gartstein, M. Litinskaya, Hybrid resonant organic-inorganic nanostructures for optoelectronic applications. Chem. Rev. **111**, 5179–5214 (2011)
32. B. Guzelturk, P.L. Hernandez Martinez, Q. Zhang, Q. Xiong, H. Sun, X.W. Sun, A.O. Govorov, H.V. Demir, Excitonics of semiconductor quantum dots and wires for lighting and displays. Laser Photonics Rev. **8**, 73 (2014)
33. I. Medintz, N. Hildebrandt (ed.), *FRET-forster resonance energy transfer: from theory to applications* (Wiley-VCH, Weinheim, 2014)
34. D. Dexter, A theory of sensitized luminescence in solids. J. Chem. Phys. **21**, 836–850 (1953)
35. B. Guzelturk, H.V. Demir, Organic-inorganic composites of semiconductor nanocrystals for efficient excitonics. J. Phys. Chem. Lett. **6**, 2206–2215 (2015)

Chapter 2
Energy Transfer Review

In this chapter, we discuss the basic concepts of excitation energy transfer, making the distinction between *radiative* and *nonradiative*, and giving a brief overview on the classical and quantum mechanical description of energy transfer [1].

2.1 Introduction

Excitation energy transfer of an excited state energy of the donor (D) to the ground state of the acceptor (A) is possible provided that the emission spectrum of the donor partially overlaps the absorption spectrum of the acceptor. This is a very important process, which occurs in a variety of situations. Excitation energy transfer can be grouped into *heterotransfer* versus *homotransfer* and *radiative* versus *nonradiative* transfer.

Energy transfer from an excited donor (D^*) to another that is chemically different, ground acceptor A, is called *heterotransfer*.

$$(D^*, A) \rightarrow (D, A^*) \tag{2.1}$$

If the donor and the acceptor are identical, then the energy transfer is *homotransfer*.

$$(D^*, D) \rightarrow (D, D^*) \tag{2.2}$$

In homotransfer, excitation transport or energy migration can occur. In such case, the process of energy transfer repeats itself such that the excitation migrates over several molecules.

© The Author(s) 2016
A. Govorov et al., *Understanding and Modeling Förster-type Resonance Energy Transfer (FRET)*, Nanoscience and Nanotechnology,
DOI 10.1007/978-981-287-378-1_2

In *radiative transfer*, a photon emitted by the donor is absorbed by the acceptor, whereas *nonradiative transfer* occurs without emission of "real" photons; it is mediated by the so-called "virtual" photons. In the following we briefly discuss the classical and quantum mechanical description for the energy transfer.

2.2 Classical Description of Energy Transfer

In classical terms, the electronic energy transfer between the donor and the acceptor is viewed as the interaction of two oscillating electric dipoles. The donor's dipole is initially in oscillation, and the acceptor's dipole is initially at rest. Because of the resonance condition, the excitation is transferred from the first dipole to the second one.

The electric field of a dipole oscillating in vacuum is given by

$$\mathbf{E}(r,t) = \frac{p(t')}{4\pi\varepsilon_0} \left\{ [3(\mathbf{n}\cdot\mathbf{d})\mathbf{n} - \mathbf{d}] \left(\frac{1}{r^3} - \frac{ik}{r^2} \right) + [(\mathbf{n}\cdot\mathbf{d})\mathbf{n} - \mathbf{d}] \frac{k^2}{r} \right\} \qquad (2.3)$$

where $p(t) = p_0 \cos(\omega t)$ is the time-dependent electric dipole moment, with amplitude p_0, $t' = t - r/c$, $\mathbf{n}$ and $\mathbf{d}$ are the unit vectors in the donor to acceptor direction and the donor's dipole moment, respectively, $k = \omega/c$ and r is the distance from the dipole [2]. The distance dependence of the electric field defines two different zones: (1) for $r \ll \lambda$ (near zone) the r^{-3} term dominates, and the angular dependence is identical to that of a static dipole, with transversal and longitudinal components; and (2) for $r \ll \lambda$ (radiative or far zone, also known as wave zone), the r^{-1} term dominates, the electric field is always perpendicular to $\mathbf{n}$ (transversal field), and the radiation corresponds to a spherical wave.

The power radiated by the dipole is [2]

$$P^0 = \frac{p_0^2 \omega^4}{12\pi\varepsilon_0 c^3} \qquad (2.4)$$

For simplicity, the acceptor can be considered as a passive absorber characterized by an absorption cross-section. Therefore, the power it absorbs, when placed at a distance r from the dipole is

$$P' = \frac{1}{2} c\varepsilon_0 E_0^2 \sigma \qquad (2.5)$$

where σ is the acceptor's cross-section and E_0^2 is the amplitude of the dipole's electric field given by (2.3). After orientation averaging, E_0^2 takes the form of

$$E_0^2 = 2\left(\frac{p_0}{4\pi\varepsilon_0}\right)^2 \left(\frac{k^4}{3r^2} + \frac{k^2}{3r^4} + \frac{1}{r^6}\right) \tag{2.6}$$

Substitution of (2.6) into (2.5) gives

$$P' = \frac{\sigma}{4\pi r^2}\left[1 + \left(\frac{\lambdabar}{r}\right)^2 + 3\left(\frac{\lambdabar}{r}\right)^4\right]P^0 \tag{2.7}$$

where $\lambdabar = \lambda/(2\pi)$ For large distances $r \gg \lambdabar$, (2.7) reduces to

$$P' = \frac{\sigma}{4\pi r^2}P^0 \tag{2.8}$$

This equation has a simple geometric interpretation and corresponds to the radiative transfer. Equation (2.8) implies that the power emitted by the donor in the presence of an acceptor is

$$P = \left\{1 + \frac{\sigma}{4\pi r^2}\left[\left(\frac{\lambdabar}{r}\right)^2 + 3\left(\frac{\lambdabar}{r}\right)^4\right]\right\}P^0 \tag{2.9}$$

When the acceptor is located in the near zone $r \gg \lambdabar$, P exceeds P^0. The reason for this lies in Eq. (2.3), which allows us to see that energy is stored in the near field. Energy periodically flows out of the source and returns without being lost. This energy does not appear in the net radiative balance, which only accounts for the small amount of energy leaked as radiation. When an acceptor is located in the near zone, the donor decay rate increases because the acceptor feeds on energy temporally deposited in the field by the donor.

On the other hand, from Eq. (2.7), the absorbed power of the acceptor, when it is in the near zone, is

$$P' = \frac{3\sigma}{64\pi^5}\left(\frac{\lambda^4}{r^6}\right)P^0 \tag{2.10}$$

This equation can be written in terms of the transfer rates k_T and the radiative rate k_r, by dividing both sides by $h\upsilon$, and relating σ to the molar absorption coefficient ε_A, which gives

$$k_T = k_r\left(\frac{3\ln 10\varepsilon_A\lambda^4}{64\pi^5 N_A n^4}\right)\frac{1}{r^6} \tag{2.11}$$

where N_A is the Avogadro's number and n is the medium refractive index, which is assumed to be non-absorbing. $k_r = Q_D/\tau_0$, here τ_0 is the donor's lifetime in the absence of acceptors and Q_D is the donor's quantum yield. Moreover, if we assume a spectral distribution for the emission wavelength, then Eq. (2.11) becomes

$$k_T = \frac{1}{\tau_0} \left(\frac{Q_D 3 \ln 10}{64\pi^5 N_A n^4} \right) \left(\frac{1}{r^6} \right) \int\limits_0^\infty F_D(\lambda)\varepsilon_A(\lambda)\lambda^4 d\lambda = \frac{1}{\tau_0} \left(\frac{R_0}{r} \right)^6 \qquad (2.12)$$

where R_0 is the Förster radius, named after Theodor Förster, who first derived Eq. (2.12) from quantum mechanical [3] and classical [4] treatments of the dipole-dipole interaction of the donor and acceptor.

2.3 Quantum Mechanical Description of Energy Transfer

The quantum mechanical treatment of the energy transfer considers that only two electrons are involved in the D-A transition, one from the donor and one from the acceptor. In such case, the properly anti-symmetrized wave-functions for the initial excited state Ψ_i (D excited, A unexcited) and the final excited state Ψ_f (D unexcited, A excited) can be written as

$$\Psi_i = \frac{1}{\sqrt{2}} [\Psi_{D^*}(1)\Psi_A(2) - \Psi_{D^*}(2)\Psi_A(1)]$$
$$\Psi_f = \frac{1}{\sqrt{2}} [\Psi_D(1)\Psi_{A^*}(2) - \Psi_D(2)\Psi_{A^*}(1)] \qquad (2.13)$$

where the number 1 and 2 refer to the two electrons involved.

The interaction matrix element describing the coupling between the initial and final state is given by

$$U = \langle \Psi_i | \hat{V} | \Psi_f \rangle \qquad (2.14)$$

where $\hat{V}$ is the perturbation part of the total Hamiltonian $\hat{H} = \hat{H}_D + \hat{H}_A + \hat{V}$. U can be written as a sum of two terms

$$U = \langle \Psi_{D^*}(1)\Psi_A(2) | \hat{V} | \Psi_D(1)\Psi_{A^*}(2) \rangle - \langle \Psi_{D^*}(1)\Psi_A(2) | \hat{V} | \Psi_D(2)\Psi_{A^*}(1) \rangle \quad (2.15)$$

The first term in Eq. (2.15) is called the *Coulombic term*, U_C, where the initially excited electron on D returns the ground state while an electron on A is simultaneously promoted to the excited state. The second term in (2.15) is called the *exchange term*, U_{ex}, here there is an exchange of electrons on D and A. The exchange interaction is a quantum mechanical effect arising from the symmetry properties of the wave-functions with respect to the exchange of spin and space coordinates of two electrons.

The Coulombic term can be expanded into a sum of multipole-multipole terms, however, it is generally approximated by the first predominant term representing the dipole-dipole interaction between the transition dipoles moments $\mathbf{M}_D$ and $\mathbf{M}_A$ of the transitions $D \rightarrow D^*$ and $A \rightarrow A^*$. It is worth mentioning that the squares of the transition dipole moments are proportional to the oscillator strengths of these transitions. Thus, the Coulombic term can be written as

$$U_{d-d} = \frac{\mathbf{M}_D \cdot \mathbf{M}_A}{r^3} - \frac{(\mathbf{M}_A \cdot \mathbf{r})(\mathbf{M}_D \cdot \mathbf{r})}{r^5} \tag{2.16}$$

where r is the donor-acceptor separation distance. This expression can be simplified into

$$U_{d-d} = 5.04 \frac{|\mathbf{M}_D||\mathbf{M}_A|}{r^3}(\cos\theta_{DA} - 3\cos\theta_D \cos\theta_A) \tag{2.17}$$

U_{d-d} is expressed in cm^{-1}, the transition moments in Debye units, and r in nanometers. θ_{DA} is the angle between the two transition moments and θ_D and θ_A are the angles between each transition moment and the vector connecting them. Note that 1 *Debye* unit is equal to 3.33×10^{-30} C m.

The dipole approximation is valid only for point-like dipoles, i.e., when the D-A separation is much larger than the donor and acceptor dimensions. At the short distances or when the dipole moments are large, higher multipole terms should be included in the calculations.

The exchange term represents the electrostatic interaction between the charge clouds. The transfer occurs via the overlap of the electron clouds and requires physical contact between the donor and the acceptor. The interaction is short range because the electron density decays exponentially outside the boundaries. For two electrons separated by the distance r_{12} in the D-A pair, the space part of the exchange interaction can be written as

$$U_{ex} = \left\langle \Phi_{D^*}(1)\Phi_A(2) \left| \frac{e^2}{r_{12}} \right| \Phi_D(2)\Phi_{A^*}(1) \right\rangle \tag{2.18}$$

where Φ_D and Φ_A are the contributions of the spatial wave-functions to the total wave-function Ψ_D and Ψ_A that include the spin functions. The spin selection rules (Wigner's rule) for allowed energy transfer are obtained by integration over the spin coordinates.

The transfer rate k_T is given by the *Fermi's Golden* rule

$$k_T = \frac{2\pi}{\hbar}|U|^2\rho \tag{2.19}$$

where ρ is a measure of the density of the interacting initial and final states, as determined by *Franck-Gordon* factors, and is related to the overlap integral between the emission spectrum of the donor and the absorption spectrum of the acceptor.

The Förster rate constant for the energy transfer in the case of long-range dipole-dipole interaction is obtained by substituting Eq. (2.17) into Eq. (2.19), whereas the Dexter rate constant for the short-range exchange interaction is obtained by substituting Eq. (2.18) into Eq. (2.19).

2.4   Radiative and Nonradiative Energy Transfer

Radiative energy transfer corresponds to the absorption of a photon, by the acceptor, upon emission by the donor and it is observed when the average distance between the donor and the acceptor is larger than the emitted photon wavelength. Such transfer does not require any interaction between D-A pairs, but it depends on the spectral overlap and the concentration. On the other hand, nonradiative energy transfer occurs at subwavelength distances and without the emission of a photon; and it is the result of the short- and long-range interactions between D-A pair. For instance, nonradiative energy transfer by the dipole-dipole interaction can extend over distances up to nearly 20 nm. Such kind of transfer provides a tool for determining separations of a few nm between D-A pair.

2.4.1   Radiative Energy Transfer

Radiative energy transfer is a two-step process where a photon emitted by the donor is absorbed by the acceptor.

$$(1) \qquad\qquad\qquad D^* \rightarrow D + h\upsilon \qquad\qquad\qquad (2.20)$$

$$(2) \qquad\qquad h\upsilon + A \rightarrow A^* \quad \text{or} \quad h\upsilon + D \rightarrow D^* \qquad\qquad (2.21)$$

This process is usually called trivial transfer because of the simplicity of the phenomenon; however the quantitative description is complicated because it depends on the sample's size and its configuration with respect to excitation and observation.

The fraction f of photons emitted by the donor and absorbed by the acceptor is given by

$$f = \frac{1}{Q_D} \int\limits_0^\infty I_D(\lambda) \left[1 - 10^{-\varepsilon_A(\lambda)C_A l} \right] d\lambda \qquad (2.22)$$

where C_A is the molar concentration of acceptors, Q_D is the donor's quantum yield in the absence of acceptor, l is the thickness of the sample, $\varepsilon_A(\lambda)$ is the molar absorption coefficient of the acceptor, and $I_D(\lambda)$ is the donor emission intensity with the normalization condition

$$Q_D = \int\limits_0^\infty I_D(\lambda)d\lambda \tag{2.23}$$

If the optical density is not too large, f can be approximated by

$$f = \frac{2.3}{Q_D}C_A l \int\limits_0^\infty I_D(\lambda)\varepsilon_A(\lambda)d\lambda \tag{2.24}$$

where the integral represents the overlap between the donor emission spectrum and the acceptor absorption spectrum.

2.4.2 Nonradiative Energy Transfer

Nonradiative energy transfer requires interaction between the donor and the acceptor. It can occur if the emission spectrum of the donor overlaps the absorption spectrum of the acceptor so that several vibronic transitions between the donor and the acceptor, with the same energy at the donor and acceptor side, couple and thus are in resonance. This type of transfer is called resonance energy transfer (RET).

There are different types of interaction that are involved in resonance energy transfer. These interactions may be Coulombic and/or due to intermolecular orbital overlap. The Coulombic interactions consist of the long-range dipole-dipole interactions (Förster's mechanism) and short-range multipolar interactions. The interaction due to intermolecular orbital overlap, which includes electron exchange (Dexter's mechanism) and charge resonance interactions, are short range. It is worth mentioning that for the singlet-singlet energy transfer ($^1D^* + {}^1A \rightarrow {}^1D + {}^1A^*$), all types of interactions are involved, whereas the triplet-triplet energy transfer ($^3D^* + {}^1A \rightarrow {}^1D + {}^3A^*$) is due to only the orbital overlap. For allowed transition between the D-A pair, the Coulomb interaction is predominant, even at short distances. For forbidden transition between the D-A, the exchange mechanism is dominant. The interaction distance range for the exchange mechanism is usually less than 1 nm; and for the Coulombic mechanism, it is usually less than 20 nm.

There are three main classes of coupling, depending on the relative values of the interaction energy (U), the electronic difference between D^* and A^* (ΔE), the absorption bandwidth (Δw), and the vibronic bandwidth ($\Delta \varepsilon$).

(1) *Strong Coupling* ($U \gg \Delta E, U \gg \Delta w, \Delta \varepsilon$): In this case, the Coulombic term U_C is much larger than the width of the individual transition $D \rightarrow D^*$ and $A \rightarrow A^*$. Thus, all the vibronic subtransitions in the D-A pair are in resonance

with one another. Here, the transfer energy is faster than the nuclear vibrations and vibrational relaxation ($\sim 10^{-12}$ s). The excitation energy is delocalized over the two components oscillating back and forth between the D and A. This type of energy transfer is a coherent process and is described in the frame of exciton theory. Strong coupled systems are characterized by large differences between their absorption spectra and those of their components. For a two-component system, two new absorption bands are observed due to transitions of the in-phase and out-of-phase combinations of the locally excited states. These two transitions are separated in energy by $2|U|$.

The transfer rate for strongly coupled system is

$$k_T \approx \frac{4|U|}{h} \tag{2.25}$$

When U is approximated by the dipole-dipole interaction, the distance dependence of U and, therefore, k_T are r^{-3}.

(2) *Weak Coupling* ($U \gg \Delta E, \Delta w \gg U \gg \Delta \varepsilon$): In this case, the interaction is much lower than the absorption bandwidth but larger than the width of an isolated vibronic level. Here, the vibronic excitation is considered as delocalized so that the system can be described in terms of stationary vibronic exciton states. This kind of system is characterized by minor alterations of the absorption spectrum.

The transfer rate is

$$k_T \approx \frac{4|U|S_{vw}^2}{h} \tag{2.26}$$

where S_{vw} is the vibrational overlap integral of the intramolecular transition $v \leftrightarrow w$. This is the transfer rate between an excited molecule with vibrational quantum number v and an unexcited one with quantum number w. Here, the transfer rate is fast compared to the vibrational relaxation but slower than the nuclear motions. Here, $S_{vw} < 1$, thus the transfer rate is slower than in the case of strong coupling.

(3) *Very Weak Coupling* ($U \ll \Delta \varepsilon \ll \Delta w$): The interaction energy is much lower than the vibronic bandwidth. Thus, the vibrational relaxation occurs before the transfer takes place. This kind of interaction does not alter the absorption spectra. The transfer rate is given by

$$k_T \approx \frac{4\pi^2 \left| US_{vw}^2 \right|^2}{h\Delta \varepsilon} \tag{2.27}$$

Here, the transfer rate depends on the square of the interaction energy as compared to the previous cases. Thus, for dipole-dipole interactions, the distance dependence is r^{-6} instead of r^{-3} for the preceding cases.

In Chap. 3 we are discussing the main feature of Förster type nonradiative energy transfer, which corresponds to the very weak coupling case. Also, in the following Chapters we will only discuss the case of very weak coupling.

References

1. B. Valeur, M.N. Berberan-Santos, *Molecular fluorescence: principles and applications*, 2nd edn. (Wiley-VCH, Weinheim, 2012)
2. M. Born, E. Wolf, *Principles of optics*, 7th edn. (Cambridge University Press, Cambridge, 1999)
3. T. Förster, Zwischenmolekulare energiewanderung und fluoreszens. Ann. Phys. **437**, 55–75 (1948)
4. T. Förster, *Fluoreszenz Organischer Verbindungen* (Vandenhoeck & Ruprecht, Göttingen, 1951)

Chapter 3
Förster-Type Nonradiative Energy Transfer Models

In this chapter, we present and discuss models for describing Förster-type nonradiative energy transfer (NRET). In the first part, we explain the main features of Förster-type NRET. In the second part, we give a brief description of another process of NRET, Dexter-type energy transfer or charge transfer. This section is reprinted (adapted) with permission of Ref. [1]. Copyright 2014 Laser and Photonics Reviews (John Wiley and Sons).

3.1 Nonradiative Energy Transfer

Energy transfer from excited donor to unexcited acceptor is a common phenomenon that occurs in nature. The excitation processes involved in energy transfer can be either radiative, or nonradiative, or both. For radiative energy transfer, a photon is emitted by an excited donor and this photon is absorbed by an unexcited acceptor. In the case of nonradiative energy transfer (NRET), energy is transmitted from the excited donor to the unexcited acceptor by a process or processes where no photon is emitted by the excited donor. This nonradiative character of the process ensures a high efficiency of NRET. One of the most important examples of nonradiative energy transfer is fluorescence resonance energy transfer, which is also known as a Förster-type resonance energy transfer (FRET). FRET is an electrodynamic phenomenon and is the result of long-range dipole-dipole interactions between the donor and the acceptor. The rate of energy transfer depends on the extent of spectral overlap of the emission spectrum of the donor with the absorption spectrum of the acceptor, the quantum yield of the donor, the relative orientation of the donor and acceptor transition dipoles, and the spatial distance between the donor and the acceptor.

© The Author(s) 2016
A. Govorov et al., *Understanding and Modeling Förster-type Resonance Energy Transfer (FRET)*, Nanoscience and Nanotechnology,
DOI 10.1007/978-981-287-378-1_3

The process of energy transfer can be described as a transition between two states:

$$(D^*, A) \xrightarrow{k_T} (D, A^*) \tag{3.1}$$

where $D^*(D)$ is the donor in the excited (ground) state, A (A^*) is the acceptor in the ground (excited) state, and k_T is the rate of resonance energy transfer (RET) between the donor and acceptor pair. In this process, the donor absorbs an external photon leaving it in an excited state. Then, the donor transfers its excited energy, via a nonradiative process, to the acceptor leaving it in an excited state.

Förster was the first to describe this process correctly [2–4]. Förster derived an expression for the resonance energy transfer and the formulation of the FRET rate and efficiency has been described in detail in various textbooks and reviews [5, 6]. From Förster's theory, the rate of energy transfer from the donor to the acceptor $k_T(r)$ is given by [6]

$$k_T(r) = \frac{1}{\tau_D} \left(\frac{R_0}{r} \right)^6 \tag{3.2}$$

where τ_D is the decay time of the donor in the absence of an acceptor, R_0 is the Förster radius, and r is the donor-to-acceptor distance. Looking at (3.2), the rate of energy transfer depends strongly on the distance and is proportional to r^{-6}. In addition, the rate of transfer is equal to the decay rate of the donor $1/\tau_D$ when r is equal to R_0, and the resulting transfer efficiency is 50 %. From this observation, we define the Förster radius as the distance at which FRET is 50 % efficient, which typically ranges from 1 to 10 nm. At this distance $(r = R_0)$, the donor emission decreases to half its intensity in the absence of acceptors.

In a more detailed study of FRET [6, 7], the rate of transfer for a single donor and a single acceptor separated by a distance r can be written as

$$k_T(r) = \frac{Q_D \kappa^2}{\tau_D r^6} \left(\frac{9000(\ln 10)}{128 \pi^5 N_A n^4} \right) \int_0^\infty F_D(\lambda) \varepsilon_A(\lambda) \lambda^4 d\lambda \tag{3.3}$$

where Q_D is the quantum yield of the donor in the absence of acceptors, n is the refractive index of the medium, N_A is Avogadro's number, r is the distance between the donor and the acceptor, and τ_D is the lifetime of the donor in the absence of acceptors. The term κ^2 is the factor describing the relative orientation of the transition dipoles of the donor and the acceptor in space. κ^2 is taken 2/3 for dynamic random averaging of the donor and the acceptor. $F_D(\lambda)$ is the normalized fluorescence intensity of the donor in the wavelength range λ to $\lambda + \Delta\lambda$ with the total intensity (area under the curve) normalized to unity. $\varepsilon_A(\lambda)$ is the extinction coefficient of the acceptor at λ, which is typically in units of M^{-1} cm^{-1}.

The overlap integral $J(\lambda)$ expresses the degree of spectral overlap between the donor emission and the acceptor absorption:

$$J(\lambda) = \int_0^\infty F_D(\lambda)\varepsilon_A(\lambda)\lambda^4 d\lambda \qquad (3.4)$$

$$J(\lambda) = \frac{\int_0^\infty F_D(\lambda)\varepsilon_A(\lambda)\lambda^4 d\lambda}{\int_0^\infty F_D(\lambda)d\lambda} \qquad (3.5)$$

$F_D(\lambda)$ is dimensionless. In calculating $J(\lambda)$ one should use the corrected emission spectrum with its area normalized to unity, or normalize the calculated value of $J(\lambda)$ by the area. The most common units of $J(\lambda)$ are: (1) $M^{-1}cm^3$, if $\varepsilon_A(\lambda)$ is expressed in units of $M^{-1}cm^{-1}$ and λ is taken in centimeters, and (2) $M^{-1}cm^{-1}nm^4$, if $\varepsilon_A(\lambda)$ is expressed in units of $M^{-1}cm^{-1}$ and λ is in nanometers $\left(M = \frac{mol}{L}\right)$.

For practical reasons it is easier to think in terms of the spatial distance rather than transfer rate. Thus, (3.2) is then expressed in terms of the Förster radius R_0. From (3.2) and (3.3) one obtains:

$$R_0^6 = \left(\frac{9000(\ln 10)Q_D\kappa^2}{128\pi^5 N_A n^4}\right) \int_0^\infty F_D(\lambda)\varepsilon_A(\lambda)\lambda^4 d\lambda \qquad (3.6)$$

This expression allows to calculate the Förster radius from the spectral properties of the donor and the acceptor and the donor quantum yield. The efficiency of energy transfer (ξ) is the fraction of photons absorbed by the donor of which excitation energy is transferred to the acceptor. This fraction is given by

$$\xi = \frac{k_T(r)}{\tau_D^{-1} + k_T(r)} \qquad (3.7)$$

which is the ratio of the transfer rate to the total decay rate of the donor in the presence of the acceptor. From (3.7), we can observe: (1) when the transfer rate is much faster than the decay rate, the energy transfer is efficient; and (2) when the transfer rate is slower than the decay rate, the energy transfer is inefficient because only a little transfer occurs during the excited state lifetime.

The efficiency of energy transfer can be written as a function of the distance by substituting (3.2) into (3.7).

Fig. 3.1 Energy transfer efficiency (ξ) versus distance normalized with respect to R_0. R_0 is the Förster radius

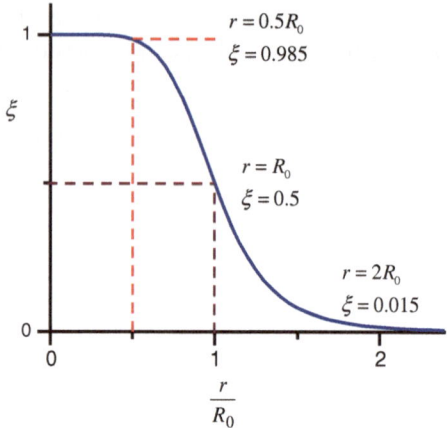

$$\xi = \frac{R_0^6}{R_0^6 + r^6} \tag{3.8}$$

This equation clearly shows that the transfer efficiency is strongly dependent on the distance when the D-A distance is near to R_0 (Fig. 3.1). The efficiency quickly increases to unity as the D-A distance decreases below R_0. Conversely, the efficiency quickly diminishes if r is greater than R_0. Note that when $r = 2R_0$, the transfer efficiency is 1.5 %, and when $r = 0.5R_0$, the transfer efficiency is 98.5 %.

FRET is a highly distance sensitive process owing to the inverse sixth power (r^{-6}) dependence of the separation distance in the case of point-to-point dipole coupling. Therefore, FRET was first used as a nanoscale ruler [8]. FRET distance dependency is altered for different acceptor geometries. For example, small molecules and 3D-confined quantum dot (QD) acceptors are considered to be infinitesimal transition dipoles, which lead to the classical r^{-6} dependence in the case of a-single-donor-to-a-single-acceptor. On the other hand, 2D confined quantum wire (NW) and 1D confined quantum well (QW) acceptors lead to distance dependences that vary with r^{-5} and r^{-4}, respectively [9, 10]. Basically, quantum confinement of the acceptor modifies the distance dependency of the FRET. Furthermore, different assemblies of the acceptors can also alter the distance dependence, as in the case where a 2D-like assembly of QDs (i.e., a monolayer of QDs on a QW donor) act as a 1D-confined structure, which consequently results in the distance dependence having the form of r^{-4} similar to QWs [9, 10]. Note that the confinement of the donor (i.e., the dimensionality of the donor) tailors the effective dielectric constant [10], which is discussed in Chap. 6 of the 2nd brief of the series.

FRET has been widely exploited in various application areas of molecular biology for sensing, labelling, nanoscale distance measurements and understanding of the molecular-level interactions. For these biological systems, typically in solution, point-to-point like interaction is effective, and, thus, r^{-6} dependence is commonly valid. Recently, FRET has been shown to be useful for optoelectronic technologies towards the purpose of creating highly efficient lighting and solar energy–harvesting systems. For this aim, exciton energy transfer in the QD-, NW-, and QW-based nanostructures can be employed to improve and control the photonic properties for light-generation and -harvesting systems. In these systems, with dimensionality of particles systems and their assemblies, one has to be careful about the distance dependence. In the following chapters, we will describe the theory of FRET in extended nanostructures (assemblies).

To sum up, here we briefly discussed FRET beyond the dipole-dipole approximation. In the case of multipole Coulomb interaction, such as dipole-quadrupole and quadrupole-quadrupole interaction, the FRET rates become proportional to r^{-8} and r^{-10}, respectively [11]. As can be seen, the energy transfer rate becomes increasingly more dependent on distance, becoming more spatially sensitive. In addition, the interaction range becomes shorter. Therefore, the dominant term is the dipole-dipole interaction term, and higher other poles may be considered for larger QDs and/or when the donor and the acceptor are in very close proximity.

3.2 Dexter Energy Transfer, Charge Transfer, Exciton Diffusion and Dissociation

Dexter-type energy transfer [12], which is also known as the charge (electron) exchange energy transfer, relies on the wavefunction overlap of the electronic states between different molecules in the near field. Dexter energy transfer is a short-range energy transfer unlike FRET, which is known to be a long-range energy transfer owing to the working distances that are on the order of 10 nm. Dexter energy transfer is only effective for the donor-acceptor separations that are typically on the order of 1 nm or shorter. Dexter energy transfer can also occur between non-emissive electronic states of the materials, such as spin-forbidden triplet states, whereas it is currently believed that these excitons cannot be transferred via FRET because they have negligible oscillator strengths [13]. These exchange mechanisms are based on the Wigner spin conservation rule; thus, the spin-allowed processes are: (1) singlet-singlet energy transfer:

$$^{1}D^{*} + {}^{1}A \xrightarrow{k_{Dexter}} {}^{1}D + {}^{1}A^{*}$$

and (2) triplet-triple energy transfer:

$$^{3}D^{*} + {}^{1}A \xrightarrow{k_{Dexter}} {}^{1}D + {}^{3}A^{*}$$

Dexter energy transfer has exponential distance dependence as compared to the $k_T \propto r^{-3} - r^{-6}$ distance dependencies in the long-range FRET processes [14]. Dexter energy transfer can be written as

$$k_{Dexter} = \frac{2\pi}{h} KJ' \exp(-2r/L) \tag{3.9}$$

$$k_{Dexter} = k_0 \exp\left[\frac{2(r - R_C)}{L}\right] \tag{3.10}$$

where J' is the integral overlap

$$J' = \int_0^\infty F_D(\lambda)\varepsilon_A(\lambda)d\lambda \tag{3.11}$$

with the normalization condition

$$\int_0^\infty F_D(\lambda)d\lambda = \int_0^\infty \varepsilon_A(\lambda)d\lambda = 1 \tag{3.12}$$

and R_C is the distance of the closest approach (collisional radius) and L is the average Bohr radius. K is a constant that is not related to any spectroscopic data. Please note that J' in (3.11) is different than J in (3.4).

Another important excitonic process is exciton diffusion. The exciton diffuses in a material in the broadened density of states of the same material; this process is called energy migration. Exciton diffusion has been widely studied for the organic semiconductors in the search for materials with large diffusion lengths to increase the probability of charge separation at the donor-acceptor hetero-interfaces in organic solar cells [15]. In addition to organic materials, exciton diffusion is crucial in bulk and quantum-confined semiconductor structures i.e., QWs, NWs, and QDs assemblies. Excitons can be transported in these quantum-confined materials; however, these systems should be well understood and controlled because defects can trap the diffusing excitons such that the emission efficiencies significantly reduce due to the increase of nonradiative recombination channels of the excitons. This picture is also valid for organic semiconductors.

Exciton dissociation is the separation of the bound electron-hole pairs into free carriers. This dissociation is a central step for excitonic solar cells [16] (bulk-heterojunction [17] and dye-sensitised [18]) because the generation of free charge carriers is required to realise the photovoltaic operation. In excitonic solar cells, dissociation of the excitons is facilitated by the interfaces that have type-II band alignments to physically break the excitons into free charges. The resistance against the disassociating of excitons in terms of energy is called the exciton binding energy. Materials with larger exciton binding energy have more stable excitons because it is difficult to overcome this large Coulomb energy between the electron-hole pairs.

Lately, excitonic processes such as multi-exciton generation (MEG), Auger recombination and exciton-exciton annihilation have been studied in the quantum-confined semiconductors. Multi-exciton generation, also dubbed carrier multiplication, is the generation of multi-excitons upon the absorption of a high-energy photon $hv \geq 2 \times E_{Gap}$. It has been shown that semiconductor QDs can be very efficient in terms of converting higher-energy photons into multi-excitons [19, 20]. Related to the multi-exciton phenomena, Auger recombination becomes severe because excitons are spatially very close to each other. In Auger recombination, the energy of the recombining exciton is transferred to another already excited charge carrier in the material such that this charge is excited into higher energy states (i.e., hot carrier). This hot carrier quickly thermalizes to the respective band edge by losing its energy to the phonon vibrations; therefore, Auger recombination can significantly decrease the multi-exciton operation in the quantum-confined structures [21].

3.3   Selection Rules for Enery Transfer

In this section we summarize the processes that are allowed under the dipole-dipole and exchange mechanisms.

3.3.1   Dipole-Dipole Mechanism

- $^1D^* + {}^1A \rightarrow {}^1D + {}^1A^*$: Singlet-Singlet Energy Transfer.
- $^1D^* + {}^3A^* \rightarrow {}^1D + {}^3A^{**}$: Higher Triplet Energy Transfer. This type of transfer requires overlap of the fluorescence spectrum of the donor and the T-T absorption spectrum of the acceptor. In this case, both donor and acceptor are in the excited states, but FRET formalism remains valid, with a few adaptations.
- $^3D^* + {}^1A \rightarrow {}^1D + {}^1A^*$: Triplet-Singlet Energy Transfer. This type of transfer leads to phosphorescence quenching of the donor.
- $^3D^* + {}^3A^* \rightarrow {}^1D + {}^1A^{**}$: Higher Triplet Energy Transfer. This type of transfer requires overlap of the phosphorescence spectrum of D^* and the T-T absorption spectrum of A. The donor and acceptor are both in excited states.

3.3.2   Exchange Mechanism

- $^1D^* + {}^1A \rightarrow {}^1D + {}^1A^*$: Singlet-Singlet Energy Transfer.
- $^3D^* + {}^1A \rightarrow {}^1D + {}^3A^*$: Triplet-Triplet Energy Transfer. This type of transfer is possible because the exchange mechanism does not imply transition moments of the donor and acceptor.
- $^3D^* + {}^3A^* \rightarrow {}^1D + {}^1A^*$: Triplet-Triplet Annihilation. This type of transfer part of the energy resulting from the annihilation allows one of the two partner to return to the singlet state from which fluorescence is emitted, but with a delay determined by the triplet state lifetime.

References

1. B. Guzelturk, P.L. Hernandez Martinez, Q. Zhang, Q. Xiong, H. Sun, X.W. Sun, A.O. Govorov, H.V. Demir, Excitonics of semiconductor quantum dots and wires for lighting and displays. Laser Photonics Rev. **8**(1), 73–93 (2014)
2. Th Förster, Zwischenmolekulare energiewanderung und fluoreszens. Ann. Phys. **437**, 55–75 (1948)
3. Th. Förster, Energieanwenderung und fluoreszenz. Naturwissenschaften **6**, 166–175 (1946)
4. Th. Förster, Expermentelle und theoretische untersuchtung des zwischengmolekularen übergangs von elektronenanregungsenergie. Z. Elektrochem. **53**, 93–100 (1949)
5. T.W.J. Gadella, Förster resonance energy transfer—FRET what is it, why do it, and how it's done by R.M.Clegg, Chap. 1, in *Laboratory techniques in biochemistry and molecular biology*, vol. 33 (Academic Press, Burlington, 2009)
6. J.R. Lakowicz, *Principles of fluorescence spectroscopy*, 3rd edn. (Springer, Berlin, 2010)
7. R.M. Clegg, Fluorescence resonance energy transfer, in *Fluorescence imaging spectroscopy and microscopy*, ed. by X.F. Wang, B. Herman (Wiley, New York, 1996), pp. 179–252
8. L. Stryer, R.P. Haugland, Energy transfer: a spectroscopic ruler. PNAS **58**, 719–726 (1967)
9. V.M. Agranovich, Y.N. Gartstein, M. Litinskaya, Hybrid resonant organic-inorganic nanostructures for optoelectronic applications. Chem. Rev. **111**, 5179–5214 (2011)
10. P.L. Hernandez Martinez, A. Govorov, H.V. Demir, J. Phys. Chem. C **117**(19), 10203–10212 (2013)
11. R. Baer, E. Rabani, Theory of resonance energy transfer involving nanocrystals: the role of high multipoles. J. Chem. Phys. **128**, 184710 (2008)
12. D.L. Dexter, A theory of sensitized luminescence in solids. J. Chem. Phys. **21**, 836 (1953)
13. A. Köhler, H. Bassler, Triplet states in organic semiconductors. Mater. Sci. Eng., R **66**, 71–109 (2009)
14. B. Valeur, *Molecular fluorescence principles and applications* (Wiley-VCH Verlag GmbH, Weinheim, 2002)
15. J.-L. Bredas, R. Silbey, Excitons surf along conjugated polymer chains. Science **323**, 348–349 (2009)
16. B.A. Gregg, Excitonic solar cells. J. Phys. Chem. B **107**, 4688–4698 (2003)
17. N.S. Saricifcti, L. Smilowitz, A.J. Heeger, F. Wudl, Photoinduced electron transfer from a conducting polymer to buckminsterfullerene. Science **258**, 1474–1476 (1992)
18. B. O'Regan, M. Gratzel, A low-cost, high efficiency solar cell based on dye-sensitized colloidal TiO_2 films. Nature **353**, 737–740 (1991)

19. A.J. Nozik, Multiple exciton generation in semiconductor quantum dots. Chem. Phys. Lett. **457**, 3–11 (2008)
20. M.C. Beard, Multiple exciton generation in semiconductor quantum dots. J. Phys. Chem. Lett. **2**, 1282–1288 (2011)
21. V.I. Klimov, A.A. Mikhailovsky, S. Xu, A. Malko, J.A. Hollignsworth, C.A. Leatherdale, H.-J. Eisler, M.G. Bawendi, Optical gain and stimulated emission in nanocrystal quantum dots. Science **290**, 314–317 (2000)

Chapter 4
Background Theory

In this chapter, we briefly introduce the quantum confinement for different geometries and the formalism of the Fermi's Golden Rule. More details can be found in any quantum mechanics book, e.g., Refs. [1, 2].

4.1 Quantum Cofinement

4.1.1 Three Dimensional Cartesian Coordinates

The time-dependent Schrödinger equation for a spinless particle of mass m moving under the influence of a three-dimensional potential is

$$-\frac{\hbar^2}{2m}\vec{\nabla}^2\Psi(x,y,z,t) + \hat{V}(x,y,z,t)\Psi(x,y,z,t) = i\hbar\frac{\partial\Psi(x,y,z,t)}{\partial t} \qquad (4.1)$$

where $\vec{\nabla}$ is the Laplacian given by

$$\vec{\nabla}^2 = \frac{\partial^2}{\partial x^2} + \frac{\partial^2}{\partial y^2} + \frac{\partial^2}{\partial z^2} \qquad (4.2)$$

The wavefunction of a particle in a time-independent potential can be written as a product of spatial and time components:

$$\Psi(x,y,z,t) = \psi(x,y,z)e^{-i\frac{Et}{\hbar}} \qquad (4.3)$$

© The Author(s) 2016
A. Govorov et al., *Understanding and Modeling Förster-type Resonance Energy Transfer (FRET)*, Nanoscience and Nanotechnology,
DOI 10.1007/978-981-287-378-1_4

where $\psi(x, y, z)$ is the solution to the time-independent Schrödinger equation:

$$-\frac{\hbar^2}{2m}\vec{\nabla}^2\psi(x, y, z) + \hat{V}(x, y, z)\psi(x, y, z) = E\psi(x, y, z) \qquad (4.4)$$

which is of the from $\hat{H}\psi = E\psi$. If the potential can be separated into the sum of three independent, one-dimensional terms

$$V(x, y, z) = V(x) + V(y) + V(z) \qquad (4.5)$$

we can solve (4.4) by the method of *separation of variables*. This method consists of separating the three-dimensional Schrödinger equation into three independent one-dimensional Schrödinger equations. Then (4.4), in conjugation with (4.5), can be written as

$$\left[\hat{H}_x + \hat{H}_y + \hat{H}_z\right]\psi(x, y, z) = E\psi(x, y, z) \qquad (4.6)$$

where $\hat{H}_\alpha$ is given by

$$H_\alpha = -\frac{\hbar^2}{2m}\frac{\partial^2}{\partial\alpha^2} + V_\alpha(\alpha) \qquad (4.7)$$

with $\alpha = x, y, z$.

As $V(x, y, z)$ separates into three independent terms, we can also write $\psi(x, y, z)$ as a product of three functions, each with a single variable:

$$\psi(x, y, z) = X(x)Y(y)Z(z) \qquad (4.8)$$

Substituting (4.8) into (4.6) and dividing it by $X(x)Y(y)Z(z)$, we obtain

$$\left[-\frac{\hbar^2}{2m}\frac{1}{X}\frac{d^2X}{dx^2} + V_x(x)\right] + \left[-\frac{\hbar^2}{2m}\frac{1}{Y}\frac{d^2Y}{dy^2} + V_y(y)\right] + \left[-\frac{\hbar^2}{2m}\frac{1}{Z}\frac{d^2Z}{dz^2} + V_z(z)\right] = E$$
$$(4.9)$$

Since each expression in the square brackets depends on only one of the variables x, y, or z, and since the sum of these three expressions is equal to a constant energy, E, each expression must then be equal to a constant such that the sum of these three constants is equal to E. For instant, the x-dependent expression is given by

$$\left[-\frac{\hbar^2}{2m}\frac{d^2}{dx^2} + V_x(x)\right]X(x) = E_x X(x) \qquad (4.10)$$

Similar equations are also applicable for the y and z coordinates, with

$$E_x + E_y + E_z = E \qquad (4.11)$$

4.1.2 The Box Potential

We begin with the rectangular box potential, which has no symmetry, and then consider the cubic potential, which displays a great deal of symmetry, since x, y, and z axes are equivalent.

4.1.2.1 The Rectangular Box Potential

Consider first the case of a spinless particle of mass m confined in a *rectangular* box of sides L_x, L_y and L_z:

$$V(x,y,z) = \begin{cases} 0 & \text{if } 0 < x < L_x, 0 < y < L_y, 0 < z < L_z \\ \infty & \text{elsewhere} \end{cases} \qquad (4.12)$$

which can be written as $V(x,y,z) = V_x(x) + V_y(y) + V_z(z)$, with

$$V_x(x) = \begin{cases} 0 & \text{if } 0 < x < L_x \\ \infty & \text{elsewhere} \end{cases} \qquad (4.13)$$

and the potential $V_y(y)$ and $V_z(z)$ have similar forms.

The wavefunction $\psi(x,y,z)$ must vanish at the walls of the box. The solutions for this potential are of the form

$$X(x) = \sqrt{\frac{2}{L_x}} \sin\left(\frac{n_x \pi}{L_x}x\right) \quad n_x = 1, 2, 3, \dots \qquad (4.14)$$

and the corresponding energy eigenvalues are

$$E_{n_x} = \frac{\hbar^2 \pi^2}{2mL_x^2} n_x^2 \qquad (4.15)$$

From these expressions we can write the normalized three-dimensional eigen-functions and their corresponding energies:

$$\psi_{n_x n_y n_z}(x,y,z) = \sqrt{\frac{8}{L_x L_y L_z}} \sin\left(\frac{n_x \pi}{L_x}x\right) \sin\left(\frac{n_y \pi}{L_y}y\right) \sin\left(\frac{n_z \pi}{L_z}z\right) \qquad (4.16)$$

$$E_{n_x n_y n_z} = \frac{\hbar^2 \pi^2}{2m} \left(\frac{n_x^2}{L_x^2} + \frac{n_y^2}{L_y^2} + \frac{n_z^2}{L_z^2} \right) \tag{4.17}$$

where $n_x, n_y, n_z = 1, 2, 3, \ldots$.

4.1.2.2 The Cubic Box Potential

Similarly to the previous case, we consider the case of a spinless particle of mass m confined in a *cubic* box of side L.

$$V(x, y, z) = \begin{cases} 0 & \text{if } 0 < x < L, \ 0 < y < L, \ 0 < z < L \\ \infty & \text{elsewhere} \end{cases} \tag{4.18}$$

Recalling the results obtained for the rectangular case, (4.16) and (4.17), the eigenfunctions and eigenenergies are:

$$\psi_{n_x n_y n_z}(x, y, z) = \sqrt{\frac{8}{L^3}} \sin\left(\frac{n_x \pi}{L} x\right) \sin\left(\frac{n_y \pi}{L} y\right) \sin\left(\frac{n_z \pi}{L} z\right) \tag{4.19}$$

$$E_{n_x n_y n_z} = \frac{\hbar^2 \pi^2}{2mL^2} \left(n_x^2 + n_y^2 + n_z^2 \right) \tag{4.20}$$

The ground state, $n_x = n_y = n_z = 1$, has energy

$$E_{111} = \frac{3\hbar^2 \pi^2}{2mL^2} \tag{4.21}$$

There is three first excited states, corresponding to the three combination of n_x, n_y and n_z, i.e., $n_x = 2, n_y = 1, n_z = 1$, or $n_x = 1, n_y = 2, n_z = 1$ or $n_x = 1, n_y = 1$, $n_z = 2$ whose squares sum to 6. The first excited state has energy

$$E_{211} = E_{121} = E_{112} = \frac{6\hbar^2 \pi^2}{2mL^2} \tag{4.22}$$

Note that each of the first excited states is characterized by different wavefunction: ψ_{211} has wavelength L along the x-axes and wavelength $2L$ along the y- and z-axes, but for ψ_{121} and ψ_{121} the shorter wavelength is along the y-axis and the z-axis, respectively.

Whenever different states have the same energy, this energy level is said to be *degenerate*. In the case above, the first excited level is three-fold degenerate. This system has degenerate levels because of the high degree of symmetry associated with the cubic shape of the box. The degeneracy would be lifted, if the sides of the box were of unequal lengths (rectangular box).

4.1.3 Three Dimensional Spherical Coordinates

In this section, we study the structure of the Schrödinger equation for a particle of mass M moving in a spherically symmetric potential

$$V(\mathbf{r}) = V(r) \tag{4.23}$$

which is also known as the *central* potential.

The time-independent Schrödinger equation for a particle of momentum $-i\hbar\nabla$ and the potential vector $\mathbf{r}$ is

$$\left[-\frac{\hbar^2}{2M}\nabla^2 + V(r) \right]\psi(\mathbf{r}) = E\psi(\mathbf{r}) \tag{4.24}$$

The Laplacian ∇_r^2 separates into a radial part and an angular part ∇_Ω^2 as follows

$$\nabla^2 = \nabla_r^2 + \frac{1}{\hbar^2 r^2}\nabla_\Omega^2 = \frac{1}{r^2}\frac{\partial}{\partial r}\left(r^2 \frac{\partial}{\partial r} \right) - \frac{1}{\hbar^2 r^2}\hat{L}^2 = \frac{1}{r}\frac{\partial^2}{\partial r^2}r - \frac{1}{\hbar^2 r^2}\hat{L}^2 \tag{4.25}$$

where $\hat{\mathbf{L}}$ is the orbital angular momentum

$$\hat{L}^2 = -\hbar^2 \left[\frac{1}{\sin\theta}\frac{\partial}{\partial\theta}\left(\sin\theta\frac{\partial}{\partial\theta} \right) + \frac{1}{\sin^2\theta}\frac{\partial^2}{\partial\varphi^2} \right] \tag{4.26}$$

In spherical coordinates, the Schrödinger takes the form of

$$\left[-\frac{\hbar^2}{2M}\frac{1}{r}\frac{\partial^2}{\partial r^2}r - \frac{1}{2Mr^2}\hat{L}^2 + V(r) \right]\psi(\mathbf{r}) = E\psi(\mathbf{r}) \tag{4.27}$$

The first term of this equation can be viewed as the radial kinetic energy

$$-\frac{\hbar^2}{2M}\frac{1}{r}\frac{\partial^2}{\partial r^2}r = \frac{\hat{P}_r^2}{2M} \tag{4.28}$$

since the radial momentum operator is given in the Hermitian form

$$\hat{P}_r = \frac{1}{2}\left[\left(\frac{\mathbf{r}}{r}\right)\cdot\hat{P} + \hat{P}\cdot\left(\frac{\mathbf{r}}{r}\right) \right] = i\hbar\left(\frac{\partial}{\partial r} + \frac{1}{r} \right) \equiv -i\hbar\frac{1}{r}\frac{\partial}{\partial r}r \tag{4.29}$$

The second term $\mathbf{L}^2/(2Mr^2)$ can be identified with the rotational kinetic energy because it is generated from the pure rotation of the particle about the origin with the momentum of inertia with respect to the origin of Mr^2. In addition, $\hat{\mathbf{L}}^2$ commute with $\hat{\mathbf{L}}_z$ and $\hat{\mathbf{H}}$ as follows

$$\left[\hat{\mathbf{H}}, \hat{\mathbf{L}}^2\right] = \left[\hat{\mathbf{H}}, \hat{\mathbf{L}}_z\right] = 0 \tag{4.30}$$

Thus $\hat{\mathbf{H}}$, $\hat{\mathbf{L}}^2$, and $\hat{\mathbf{L}}_z$ have common eigenfunctions. The simultaneous eigen-function of $\hat{\mathbf{L}}^2$ and $\hat{\mathbf{L}}_z$ are given by the spherical harmonics

$$\hat{\mathbf{L}}^2 Y_{lm}(\theta, \varphi) = l(l+1)\hbar^2 Y_{lm}(\theta, \varphi) \tag{4.31}$$

$$\hat{\mathbf{L}}_z Y_{lm}(\theta, \varphi) = m\hbar Y_{lm}(\theta, \varphi) \tag{4.32}$$

The Hamiltonian in (4.27) is a sum of a radial part and an angular part. Thus, we can look for solutions that are products of a radial part and an angular part

$$\psi(\mathbf{r}) = \langle r \mid nlm \rangle = \psi_{nlm}(r, \theta, \varphi) = R_{nl}(r)Y_{lm}(\theta, \varphi) \tag{4.33}$$

The radial wavefunction $R_{nl}(r)$ has to be found. The quantum number n is introduced to identify the eigenvalues of $\hat{\mathbf{H}}$:

$$\hat{\mathbf{H}}|nlm\rangle = E_n|nlm\rangle \tag{4.34}$$

Substituting (4.34) into (4.27) and using the fact that $\psi_{nlm}(r, \theta, \varphi)$ is an eigen-function of $\hat{\mathbf{L}}^2$ (4.31), then dividing by $R_{nl}(r)Y_{lm}(\theta, \varphi)$ and multiplying by $2Mr^2$, we obtain an expression where the radial and angular degrees of freedom are separated into

$$\left[-\hbar^2 \frac{r}{R_{nl}} \frac{\partial^2}{\partial r^2}(rR_{nl}) + 2Mr^2(V(r) - E)\right] + \left[\frac{\hat{\mathbf{L}}^2 Y_{lm}(\theta, \varphi)}{Y_{lm}(\theta, \varphi)}\right] = 0$$

$$\left[-\hbar^2 \frac{r}{R_{nl}} \frac{\partial^2}{\partial r^2}(rR_{nl}) + 2Mr^2(V(r) - E)\right] + \left[\frac{l(l+1)\hbar^2 Y_{lm}(\theta, \varphi)}{Y_{lm}(\theta, \varphi)}\right] = 0 \tag{4.35}$$

$$\left[-\hbar^2 \frac{r}{R_{nl}} \frac{\partial^2}{\partial r^2}(rR_{nl}) + 2Mr^2(V(r) - E)\right] + \left[l(l+1)\hbar^2\right] = 0$$

The last expression only depends on r and the final expression is thus simplified as

$$-\frac{\hbar^2}{2M} \frac{d^2}{dr^2}(rR_{nl}(r)) + \left[V(r) + \frac{l(l+1)\hbar^2}{2Mr^2}\right](rR_{nl}(r)) = E_n(rR_{nl}(r)) \tag{4.36}$$

Note that (4.36) does not depend on the azimuthal quantum number m. Thus, the energy E_n is $2l(l+1)$—fold *degenerate*. This is due to the fact that, for a given l, there are $(2l+1)$ different eigenfunctions ψ_{nlm} (i.e., ψ_{nl-l}, ψ_{nl-l+1}, $\cdots$, ψ_{nll-1}, ψ_{nll}), which correspond to the same eigenenergy E_n. This degeneracy property is peculiar to the central potentials. Moreover, (4.36) has the structure of a one-dimensional equation for r as follows

$$-\frac{\hbar^2}{2M}\frac{d^2 U_{nl}(r)}{dr^2} + V_{eff}(r)U_{nl}(r) = E_n U_{nl}(r) \tag{4.37}$$

whose solutions give the energy levels of the system with the wavefunction given by

$$U_{nl}(r) = rR_{nl}(r) \tag{4.38}$$

and the potential by

$$V_{eff}(r) = V(r) + \frac{l(l+1)\hbar^2}{2Mr^2} \tag{4.39}$$

which is known as the *effective* or *centrifugal* potential. Here, $V(r)$ is the central potential and $l(l+1)\hbar^2/(2Mr^2)$ is a repulsive or centrifugal potential, which is associated with the orbital angular momentum and tends to repel the particle away from the center. $\psi_{nlm}(r,\theta,\varphi)$ is finite for all values of r spanning from 0 to ∞. Thus, if $R_{nl}(0)$ is finite, $rR_{nl}(r)$ must vanish at $r = 0$, i.e.,

$$\lim_{r\to 0}[rR_{nl}(r)] = U_{nl}(0) = 0 \tag{4.40}$$

4.1.3.1 Free Particle in Spherical Coordinates

Here, we apply the formalism developed above to study the motion of a free particle of mass M and energy $E_k = \hbar^2 k^2/(2M)$ where k is the wave vector $k = |\mathbf{k}|$. The Hamiltonian $H = -\hbar^2\nabla^2/(2M)$ of a free particle is *rotational invariant* and commutes with $\hat{\mathbf{L}}^2$ and $\hat{L}_z$. Thus, the radial equation for a free particle is

$$-\frac{\hbar^2}{2M}\frac{1}{r}\frac{d^2}{dr^2}(rR_{kl}(r)) + \frac{l(l+1)}{2Mr^2}R_{kl}(r) = ER_{kl}(r) \tag{4.41}$$

using a change of variable $\rho = kr$, we reduce this equation into

$$\frac{d^2 R_l(\rho)}{d\rho^2} + \frac{2}{\rho}\frac{dR_l(\rho)}{d\rho} + \left[1 - \frac{l(l+1)}{\rho^2}\right]R_l(\rho) = 0 \tag{4.42}$$

where $R_l(\rho) = R_l(kr) = R_{kl}(r)$. This differential equation is known as the *spherical Bessel* equation. The general solutions to this equation are given by the spherical Bessel functions $j_l(\rho)$ and the spherical Neumann functions $n_l(\rho)$. Since the Neumann functions $n_l(\rho)$ diverge at the origin, and since the wavefunctions $\psi_{klm}(r, \theta, \varphi)$ are required to be finite everywhere in space, only the spherical Bessel functions $j_l(\rho)$ contribute to the eigenfunctions of the free particle

$$\psi_{klm}(r, \theta, \varphi) = j_l(kr) Y_{lm}(\theta, \varphi) \tag{4.43}$$

where $k = \sqrt{2ME_k}/\hbar$. Note that, since the index k in $E_k = \hbar^2 k^2/(2M)$ varies *continuously*, the energy spectrum of a free particle is *infinitely degenerate*. This is because all orientations of **k** in space correspond to the same energy.

4.1.3.2   The Spherical Well

We consider a particle of mass M confined to the interior of a spherical well with impenetrable walls. In the domain $r \geq a$, the wavefunction vanishes. In the domain $r < a$, the potential is zero.

$$V(r) = \begin{cases} 0 & \text{if } r < a \\ \infty & \text{if } r \geq a \end{cases} \tag{4.44}$$

To impose the boundary condition $\psi(r = a) = 0$, we set

$$j_l(ka) = j_l(x) = j_l(x_{nl}) = 0 \tag{4.45}$$

where $x \equiv ka$ and x_{nl} is the nth zero of $j_l(x)$. Thus, the eigenfunctions and eigenvalues for the spherical well are given by nth

$$\psi_{nlm}(r, \theta, \varphi) = j_l\left(\frac{x_{nl}}{a} r\right) Y_{lm}(\theta, \varphi) \tag{4.46}$$

$$E_{nl} = \frac{\hbar^2 x_{nl}^2}{2Ma^2} \tag{4.47}$$

4.1.4   *Three Dimensional Cylindrical Coordinates*

We next consider the case of a particle of mass M confined to a cylindrical box of radius a and length L; that is,

$$V(r, \phi, z) = \begin{cases} 0 & r < a \quad 0 < z < L \\ \infty & \text{elsewhere} \end{cases} \tag{4.48}$$

Employing the Hamiltonian in cylindrical coordinates, the Schrödinger equation for the confined particle is given by

$$\frac{\partial^2 \psi}{\partial r^2} + \frac{1}{r}\frac{\partial \psi}{\partial r} + \frac{1}{r^2}\frac{\partial^2 \psi}{\partial \phi^2} + \frac{\partial^2 \psi}{\partial z^2} + k^2 \psi = 0 \tag{4.49}$$

$$E = \frac{\hbar^2 k^2}{2M} \tag{4.50}$$

with the separation of coordinates

$$\psi(r, \phi, z) = R(r)\Phi(\phi)Z(z) \tag{4.51}$$

Eq. (4.49) becomes

$$\frac{1}{R}\left(\frac{\partial^2 R}{\partial r^2} + \frac{1}{r}\frac{\partial R}{\partial r}\right) + \frac{1}{r^2}\frac{1}{\Phi}\frac{\partial^2 \Phi}{\partial \phi^2} + \frac{1}{Z}\frac{\partial^2 Z}{\partial z^2} + k^2 = 0 \tag{4.52}$$

It follows that

$$\frac{1}{Z}\frac{\partial^2 Z}{\partial z^2} = \text{constant} \equiv -k_z^2 \tag{4.53}$$

$$\frac{1}{\Phi}\frac{\partial^2 \Phi}{\partial \phi^2} = \text{constant} = -m^2 \tag{4.54}$$

$$\frac{1}{R}\left(r^2\frac{\partial^2 R}{\partial r^2} + r\frac{\partial R}{\partial r}\right) + r^2\left(k^2 - k_z^2\right) = m^2 \tag{4.55}$$

Applying boundary conditions (4.48), we find

$$Z(z) = A\sin(k_z z); k_z L = n_z \pi; n_z = 1, 2, \ldots \tag{4.56}$$

Furthermore, as $\Phi(\phi) = \Phi(\phi + 2\pi)$, we obtain

$$\Phi(\phi) = Be^{im\phi}; m = 0, \pm 1, \pm 2, \ldots \tag{4.57}$$

where A and B are constant. Returning to (4.55) and labelling

$$k^2 - k_z^2 \equiv K^2 \tag{4.58}$$

$$\rho \equiv Kr \tag{4.59}$$

There results

$$\rho^2 \frac{d^2R}{d\rho^2} + \rho \frac{dR}{d\rho} + (\rho^2 - m^2)R = 0 \tag{4.60}$$

which is known as *Bessel's equation*. General solution to this equation are given by

$$R(\rho) = C_1 J_m(\rho) + C_2 N_m(\rho) \tag{4.61}$$

where C_1 and C_2 are constants. The functions $J_m(\rho)$ and $N_m(\rho)$ are called *Bessel* and *Neumann functions of the first kind*, respectively. Since $N_m(0) = -\infty$, the only acceptable solution for the wavefunction are $J_m(\rho)$. The remaining boundary conditions gives

$$R(r = a) = C_1 J_m(aK) = 0 \tag{4.62}$$

Let us call the sth finite zero of $J_m(\rho), x_{ms}$ so that

$$J_m(aK_{ms}) \equiv J_m(x_{ms}) = 0 \tag{4.63}$$

Thus, the eigenfunctions are given by

$$\psi_{msn_z}(r, \phi, z) = A J_m \left(\frac{x_{mx}}{a} r\right) \sin\left(\frac{n_z \pi}{L} z\right) e^{im\phi} \tag{4.64}$$

with $m \geq 0$, $s > 0$, $n_z \geq 1$, and all three parameters are integers. The corresponding eigenenergies are

$$E = \frac{\hbar^2 k^2}{2M} = \frac{\hbar^2}{2M}(K^2 + k_z^2) = \frac{\hbar^2}{2M}\left[K_{ms}^2 + \left(\frac{n_z \pi}{L}\right)^2\right] \tag{4.65}$$

with $x_{ms} = aK_{ms}$, the former becomes

$$E_{msn_z} = \frac{\hbar^2}{2M}\left[\frac{x_{ms}^2}{a^2} + \left(\frac{n_z \pi}{L}\right)^2\right] \tag{4.66}$$

4.2 Fermi's Golden Rule

The transition probability corresponding to a transition from an initial unperturbed state $|\psi_i\rangle$ to another unperturbed state $|\psi_f\rangle$ is

$$P_{if}(t) = \left| -\frac{i}{\hbar} \int_0^t \langle \psi_f | \hat{V}(t') | \psi_i \rangle e^{i\omega_{fi}t'} dt' \right|^2 \tag{4.67}$$

In the case where $\hat{V}$ does not depend on time, (4.67) reduces to

$$P_{if}(t) = \left|\langle\psi_f|\hat{V}|\psi_i\rangle\right|^2 \frac{\sin^2\left(\frac{1}{2}\hbar\omega_{fi}t\right)}{\left(\frac{1}{2}\hbar\omega_{fi}\right)^2} \qquad (4.68)$$

As a function of the time, this transition probability is an oscillating sinusoidal function with a period of $2\pi/\omega_{fi}$. As a function of ω_{fi}, the transition probability, as shown in (4.68), has an interference pattern: it is appreciable only near $\omega_{fi} \simeq 0$ and decays rapidly as ω_{fi} moves away from zero (Fig. 4.1). Hence, for a fix t, we have assumed that ω_{fi} is a continuous variable; that is, we have considered a continuum of final states. This means that the transition probability of finding the system in a state $|\psi_f\rangle$ of energy E_f is the greatest only when $\omega_{fi} \simeq 0$ or equivalently when $E_i \simeq E_f$. The height and the width of the main peak, centered around $\omega_{fi} = 0$, are proportional to t^2 and $1/t$, respectively. Therefore, the area under the probability curve is proportional to t. Since most of the area is under the central peak, the transition probability is proportional to t. Therefore, the transition probability grows linearly with time. The central peak becomes narrower and stronger as time increases; this is exactly the property of a delta function. Thus, in the limit $t \to \infty$, the transition probability takes the shape of a delta function. Therefore, (4.68) boils down to

$$P_{if}(t) = \frac{2\pi t}{\hbar}\left|\langle\psi_f|\hat{V}|\psi_i\rangle\right|^2 \delta(E_f - E_i) \qquad (4.69)$$

Fig. 4.1 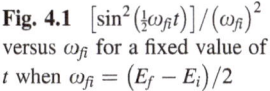 versus ω_{fi} for a fixed value of t when $\omega_{fi} = (E_f - E_i)/2$

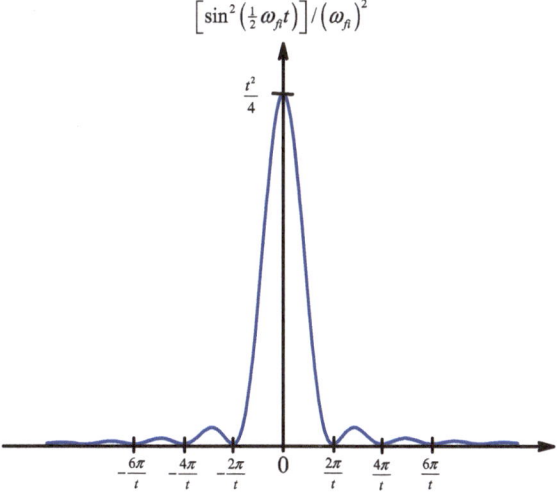

To obtain (4.69), we use the following properties:

$$\frac{\sin^2\left(\frac{1}{2}\omega_{fi}t\right)}{\left(\frac{1}{2}\omega_{fi}\right)^2} = 2\pi t\hbar\delta\left(\hbar\omega_{fi}\right) \tag{4.70}$$

$$\hbar\omega_{fi} = E_f - E_i \tag{4.71}$$

The *transition rate*, which is defined as the transition probability per unit time, is given by

$$\Gamma_{if} = \frac{P_{if}(t)}{t} = \frac{2\pi}{\hbar}\left|\langle\psi_f|\hat{V}|\psi_i\rangle\right|^2\delta\left(E_f - E_i\right) \tag{4.72}$$

The delta term $\delta\left(E_f - E_i\right)$ guarantees the conservation of energy: in the limit $t \to \infty$, the transition rate is nonvanishing only between the states of equal energy. Hence, a constant perturbation neither removes energy from the system nor supplies energy to it.

Let us now calculate the total transition rate associated with a transition from an initial state $|\psi_i\rangle$ into a continuum of final states $|\psi_f\rangle$. If $\rho\left(E_f\right)$ is the density of final states—the number of states per unit energy intervals—the number of final states within the energy intervals E_f and $E_f + dE_f$ is equal to $\rho\left(E_f\right)dE_f$. Then, the total transition rate W_{if} can be obtained from (4.72):

$$W_{if} = \int\frac{P_{if}(t)}{t}\rho\left(E_f\right)dE_f = \frac{2\pi}{\hbar}\left|\langle\psi_f|\hat{V}|\psi_i\rangle\right|^2\int\rho\left(E_f\right)\delta\left(E_f - E_i\right)dE_f \tag{4.73}$$

$$W_{if} = \frac{2\pi}{\hbar}\left|\langle\psi_f|\hat{V}|\psi_i\rangle\right|^2\rho(E_i) \tag{4.74}$$

This relation is called the *Fermi's Golden Rule*. It implies that, in the case of a constant perturbation, if we wait long enough, the total transition rate becomes constant.

References

1. R.L. Liboff, *Introductory to quantum mechanics*, 4th edn. (Addison-Wesley, San Francisco, 2003)
2. N. Zettili, *Quantum mechanics: concepts and applications* (Wiley, England, 2001)

Chapter 5
Theoretical Approaches: Exciton Theory, Coulomb Interactions and Fluctuation-Dissipation Theorem

In this chapter, we introduce the main framework for Förster-type nonradiative energy transfer; starting from exciton theory, going through Coulomb interaction, and finalizing with the fluctuation-dissipation theorem. Part of this chapter is reprinted (adapted) with permission from Ref. [1]. Copyright 2013 American Physical Society.

5.1 Electron-Hole Interaction (Exciton)

An exciton is a quasiparticle consisting of a bound state of an electron and a hole interacting via Coulomb force. An exciton can move through the medium (e.g., semiconductor crystal) and transport energy; and since an exciton is electrically neutral, it does not transport charge. An exciton can be created by external excitation, for example, through the absorption of a photon, with $E \geq E_g$. In this direct process, an electron is excited from the valence band to the conductive band, leaving behind a hole with opposite charge in the valence band, to which the electron can bind due to the attractive Coulombic interaction. Because of the Coulombic attraction between the electron and the hole in an exciton, the internal states are analogous to those of the hydrogen atom, and some of the lower energy states lie below the conduction band by an energy equivalent to the exciton binding energy in that state (Figs. 5.1 and 5.2).

An exciton has two quantities: (1) the pseudomomentum of the electron-hole pair and (2) the relative momentum of the electron and the hole. The pseudomomentum, which is equal to the vector sum of the individual momenta of the electron and the hole, enables an exciton to move throughout a crystal; and the relative momentum

© The Author(s) 2016   41
A. Govorov et al., *Understanding and Modeling Förster-type Resonance
Energy Transfer (FRET)*, Nanoscience and Nanotechnology,
DOI 10.1007/978-981-287-378-1_5

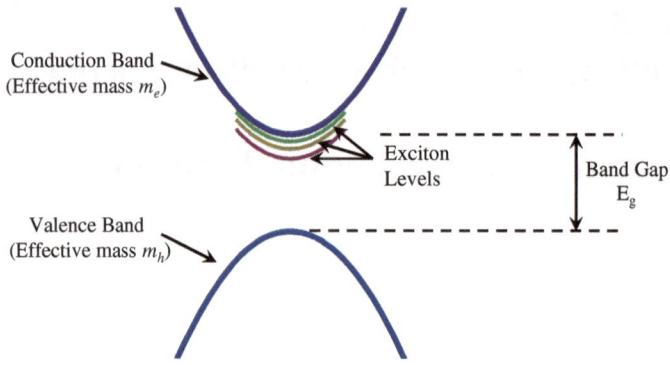

Fig. 5.1 Exciton levels for a simple band structure at $k = 0$

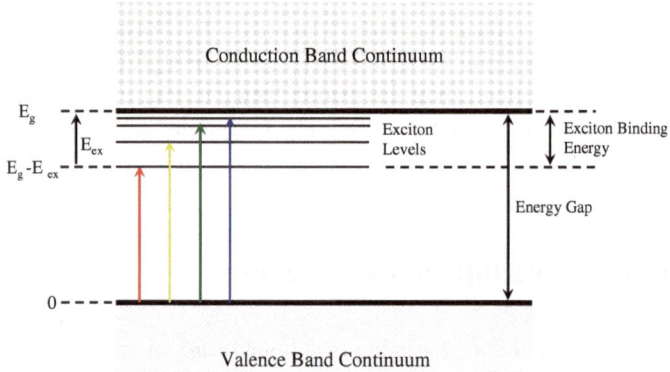

Fig. 5.2 Energy levels of an exciton created in a direct process. Optical transitions are shown by *arrows*

determines its internal structure. Excitons are classified into (1) a tightly bound exciton (Frenkel exciton) and (2) a weakly bound exciton (Mott-Wannier exciton).

5.1.1   Frenkel Excitons

In a tightly bound exciton the excitation is localized on a single atom (Fig. 5.3), i.e., a Frenkel exciton is an excited state of a single atom. A Frenkel exciton can hop from one atom to another via coupling between neighbors. Similar to all other excitation in a periodic structure, the translational states of Frenkel excitons take the form of propagating waves.

Fig. 5.3 Schematic illustration of a tightly-bound exciton (Frenkel exciton) localized on one atom in a crystal

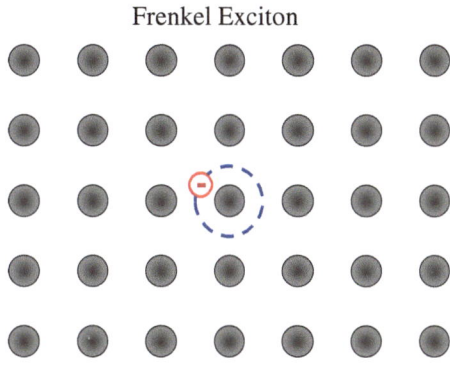

Frenkel Exciton

Consider a crystal of N atoms on a line or ring. If u_j is the ground state of atom j, the ground state of the crystal is [2]

$$\psi_g = u_1 u_2 \cdots u_j \cdots u_{N-1} u_N \tag{5.1}$$

If a single atom j is in an excited state v_j, the system is described by

$$\phi_j = u_1 u_2 \cdots u_{j-1} v_j u_{j+1} \cdots u_{N-1} u_N \tag{5.2}$$

If we consider that the excited atom interacts only with nearby atoms in its ground state, then the excitation will be passed from atom to atom.

Applying the Hamiltonian of the system on the function ϕ_j with the jth atom excited, we obtain the following

$$H\phi_j = \varepsilon\phi_j + T\left(\phi_{j-1} + \phi_{j+1}\right) \tag{5.3}$$

where ε is the free atom excitation energy; T is the transfer rate of the excitation from j to its nearest neighbors, $j - 1$ and $j + 1$. The solutions of (5.3) are the waves of the Bloch form:

$$\psi_k = \sum_j \exp(ijka)\phi_j \tag{5.4}$$

Operating the Hamiltonian on (5.4)

$$H\psi_k = \sum_j e^{ijka} H\phi_j = \sum_j e^{ijka}\left[\varepsilon\phi_j + T\left(\phi_{j-1} + \phi_{j+1}\right)\right] \tag{5.5}$$

Rearranging the right-hand side of (5.5)

$$H\psi_k = \sum_j e^{ijka}\left[\varepsilon + T\left(e^{ika} + e^{-ika}\right)\right]\phi_j = (\varepsilon + 2T\cos(ka))\psi_k \qquad (5.6)$$

so that the energy eigenvalues are (Fig. 5.4):

$$E_k = \varepsilon + 2T\cos(ka) \qquad (5.7)$$

Applying the periodic boundary conditions, the allowed values of the wavevector k are:

$$k = \frac{2\pi n}{Na}; \quad n = -\frac{1}{2}N, -\frac{1}{2}N+1, \ldots, \frac{1}{2}N - 1 \qquad (5.8)$$

5.1.2   Mott-Wannier Excitons

In a weakly bound exciton, the electron-hole distance is larger than the lattice constant of the crystal, meaning that the exciton is delocalized over several atoms (Fig. 5.5). The Mott-Wannier exciton is similar to the hydrogen atom problem. In other words, the Mott-Wannier exciton can be treated as a two-particle system weakly interacting, in which case the electron and hole energy (at $\mathbf{k} = 0$) is given by [3, 4]

$$\varepsilon_c(\mathbf{k}) = \varepsilon_c(0) + \frac{\hbar^2 k^2}{2m_e^*} \qquad (5.9)$$

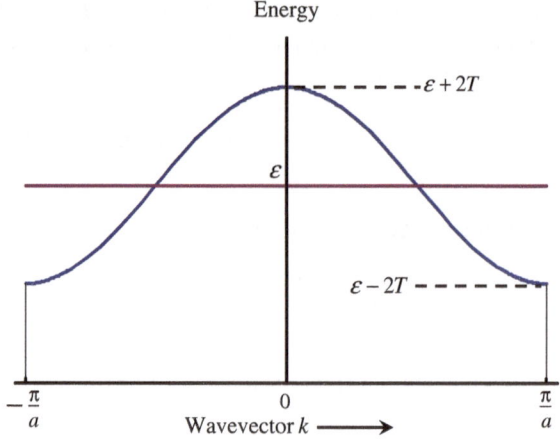

Fig. 5.4 E–k diagram for a Frenkel exciton

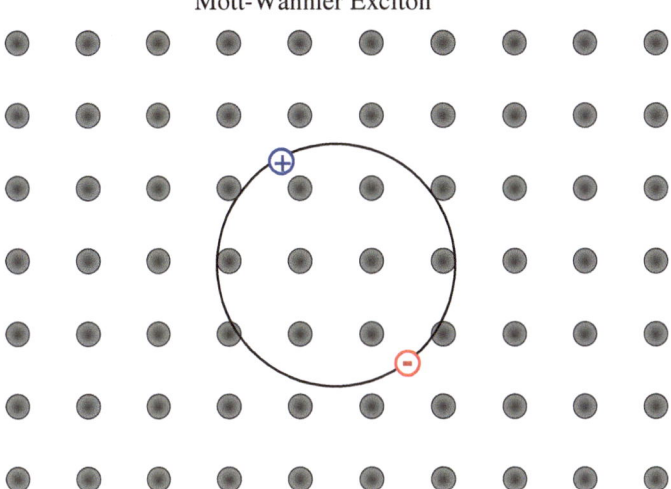

Fig. 5.5 Schematic illustration of a weakly bound exciton (Mott-Wannier exciton) delocalized over several atoms in the crystal

And

$$\varepsilon_v(\mathbf{k}) = \varepsilon_v(0) - \frac{\hbar^2 k^2}{2m_h^*} \tag{5.10}$$

where m_e^* and $-|e|$ are the electron mass and charge; and m_h^* and $+|e|$ are the hole mass and charge, respectively. For simplicity, we assume that the crystal has simple valence and conduction bands.

The total kinetic energy is

$$\mathbf{P} = \frac{\mathbf{p}_e^2}{2m_e^*} + \frac{\mathbf{p}_h^2}{2m_h^*} \tag{5.11}$$

where $\mathbf{p}_e^2$ and $\mathbf{p}_h^2$ are the electron and hole momenta, respectively. The effective Hamiltonian for the two-particle system when interacting in a dielectric medium of relative dielectric constant ε is

$$H_{\mathit{eff}} = -\frac{\hbar^2}{2m_e^*}\nabla_e^2 - \frac{\hbar^2}{2m_h^*}\nabla_h^2 - \frac{1}{4\pi\varepsilon_0\varepsilon}\frac{e^2}{|\mathbf{r}_e - \mathbf{r}_h|} \tag{5.12}$$

The solution for this Hamiltonian is

$$E_n = E_g - \frac{1}{(4\pi\varepsilon_0)^2} \frac{\mu_{ex} e^4}{2\hbar^2 \varepsilon^2} \frac{1}{n^2} + \frac{\hbar^2 \mathbf{K}^2}{2M^*} \tag{5.13}$$

where E_n is the exciton energy, $E_g = \varepsilon_c(0) - \varepsilon_v(0)$ is the bandgap energy, $\frac{1}{\mu_{ex}} = \frac{1}{m_e^*} + \frac{1}{m_h^*}$ is the *reduced* exciton mass, and $M^* = m_e^* + m_h^*$ is the *effective* exciton mass. A useful parameter for an exciton is the exciton Bohr radius (a_{ex}). It is obtained from the second term of (5.13). Therefore, the exciton Bohr is given by

$$a_{ex} = 4\pi\varepsilon_0 \frac{\hbar^2 \varepsilon}{\mu_{ex} e^2} n^2 \tag{5.14}$$

5.2   Coulombic Interaction

In both cases, the Coulombic interaction between the electron and the hole is treated with standard second order perturbation [5].

$$E = E_0 + \lambda\langle 0|H'|0\rangle + \lambda^2 \sum_i \frac{|\langle 0|H'|i\rangle|^2}{E_0 - E_i} \tag{5.15}$$

where E_0 and $|0\rangle$ are the unperturbed eigenenergy and eigenvector of the e-h ground state based on a kinetic energy, H' is the perturbation Hamiltonian, E_i and $|i\rangle$ are the unperturbed eigenenergy and eigenvector of all the other states. The effective Coulombic interaction is given by

$$H'(\mathbf{r}_e, \mathbf{r}_h) = \lambda^{-1} V(\mathbf{r}_\alpha, \mathbf{r}_\beta) \tag{5.16}$$

where $V(\mathbf{r}_\alpha, \mathbf{r}_\beta)$ is the potential function, which depends on R_{NC}, overall nanocrystal (NC) radius, and $\varepsilon = \frac{\varepsilon_{NC}}{\varepsilon_M}$ with ε_{NC} and ε_M being the NC and surrounding medium dielectric constants, respectively.

$$V(\mathbf{r}_\alpha, \mathbf{r}_\beta) = \frac{1}{4\pi\varepsilon_0\varepsilon_{NC}} \sum_{\alpha,\beta} q_\alpha q_\beta \left\{ \frac{1}{|\mathbf{r}_\alpha - \mathbf{r}_\beta|} + \frac{(\varepsilon - 1)}{R_{NC}} \right.$$
$$\left. \left[\sum_{i=1}^{\infty} \frac{(r_\alpha r_\beta)^i}{1 + \varepsilon\left(\frac{i}{i+1}\right)} P_l\left(\frac{\mathbf{r}_\alpha \cdot \mathbf{r}_\beta}{r_\alpha r_\beta}\right) + \frac{1}{2} \sum_{j=1}^{\infty} \frac{(r_\alpha)^{2j} + (r_\beta)^{2j}}{1 + \varepsilon\left(\frac{j}{j+1}\right)} \right] \right\} \tag{5.17}$$

Thus, E_X energy is given by

$$E_X = E_0^e + E_0^h + \lambda\langle 0,0|H'|0,0\rangle + \lambda^2 \sum_{\alpha,\beta} \frac{|\langle 0,0|H'|\alpha,\beta\rangle|^2}{\left(E_\alpha^e - E_0^e\right) + \left(E_\beta^h - E_0^h\right)} \qquad (5.18)$$

$$E_X = E_0^e + E_0^h - \lambda V_{00} - \lambda^2 \sum_{\alpha,\beta} \frac{|V_{\alpha\beta}|^2}{\left(E_\alpha^e - E_0^e\right) + \left(E_\beta^h - E_0^h\right)} \qquad (5.19)$$

with $\alpha \neq 0$ and $\beta \neq 0$ and $V_{\alpha\beta}$ is defined as

$$V_{\alpha\beta} = \langle 0,0|H'|\alpha,\beta\rangle \qquad (5.20)$$

5.3   Exciton in Quantum Dots: Single-Particle Quantization Energy and Coulomb Interaction

The aim is to determine for the particle in a spherical box problem the envelope wavefunction ψ for electron and hole. We consider a two-band (valance and conduction) system (Fig. 5.6). The eigenfunctions of the hole and electron are written as a product of an envelope function $\varphi_{e,h}(\mathbf{r})$ and a lattice periodic function $u_{V,C}(\mathbf{r})$ [6]:

$$\psi_e(\mathbf{r}) = \varphi_e(\mathbf{r})u_C(\mathbf{r}) \qquad (5.21)$$

$$\psi_h(\mathbf{r}) = \varphi_h(\mathbf{r})u_V(\mathbf{r}) \qquad (5.22)$$

Fig. 5.6 Quantum dot energy levels for the electron and hole in the conduction and valance band, respectively

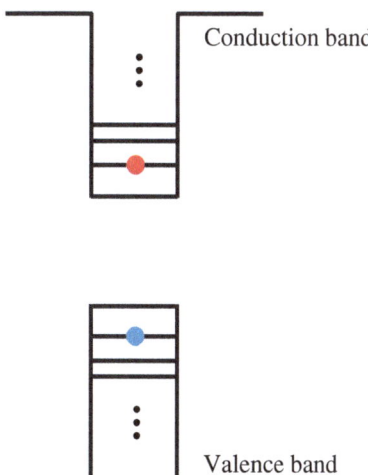

The envelope functions are the zero order eigenfunctions of the electron-hole pair Hamiltonian

$$H_{eff} = -\frac{\hbar^2}{2m_e^*}\nabla_e^2 - \frac{\hbar^2}{2m_h^*}\nabla_h^2 + V_e(\mathbf{r}_e) + V_h(\mathbf{r}_h) - \frac{1}{\varepsilon_{QD}}\frac{e^2}{|\mathbf{r}_e - \mathbf{r}_h|} \tag{5.23}$$

where m_e^*, m_h^* is the electron and home effective mass, respectively; ε_{QD} is the static dielectric constant of the nanocrystals (quantum dot); and the confinement potentials are given as follows

$$V_{e(h)}\left(r_{e(h)}\right) = \begin{cases} 0 & \text{for } r_{e(h)} < R_{QD} \\ \infty & \text{for } r_{e(h)} > R_{QD} \end{cases} \tag{5.24}$$

Here R_{QD} is the quantum dot radius.

Note that the last term on the right hand side of (5.23) is the first order perturbation term. The orthonormal functions are zero outside of the dot, and inside it are given by

$$\varphi_{nlm}^{e(h)}\left(\mathbf{r}_{e(h)}\right) = \sqrt{\frac{2}{R_{QD}^3}}\frac{j_l\left(\chi_{nl}\frac{r}{R_{QD}}\right)}{j_{l+1}(\chi_{nl})}Y_{lm}(\theta, \phi) \tag{5.25}$$

where $Y_{lm}(\theta, \phi)$ is the spherical harmonics functions, $j_l(x)$ is the *spherical Bessel* functions, and χ_{nl} are the spherical Bessel functions nth-order zeros. The energy eigenvalues $E_{nl,n'l'}$, including the requirement that the wavefunction vanishes at $r = R_{QD}$ and the first order perturbation term, are given by

$$E_{nl,n'l'} = E_g + \frac{\hbar^2}{2m_e^*}\left(\frac{\chi_{nl}^2}{R_{QD}^2}\right) + \frac{\hbar^2}{2m_h^*}\left(\frac{\chi_{n'l'}^2}{R_{QD}^2}\right) - \frac{1.8e^2}{\varepsilon_{QD}R_{QD}} \tag{5.26}$$

Here, E_g is the bulk bandgap.

5.4 Fermi's Golden Rule and Fluctuation Dissipation Theorem

In this section, we outline a macroscopic approach to the problem of dipole-dipole energy transfer. We restrict ourselves to the case of a single electron-hole pair (exciton) in the donor nanostructure. Moreover, we consider only two states ($|0\rangle$— the ground state and $|exc\rangle$—the excited state). These states are constructed using simplified wavefunctions, i.e., we consider excitonic states without mixing of the heavy- and light-hole states. Furthermore, the spin part is not included in our model.

FRET is a directional process initiated by an absorbed photon in a donor that creates an exciton in a higher excited state, relaxing very fast to the first excited state by higher order processes. This exciton can subsequently be either recombined (through radiative or nonradiative means) or transferred to an acceptor because of the Coulomb interaction between dipoles in the D-A pair. If the exciton is transferred, it will occupy a higher excited state in the acceptor and relax (very fast) to its first excited state to finally recombine through a radiative or nonradiative process. Note that FRET occurs only when the donor possesses a greater or equal bandgap compared to the acceptor. Figure 5.7 shows the energy diagram for this process.

The probability of an exciton transfer from the donor to the acceptor is given by the Fermi's Golden Rule (5.27).

$$\gamma_{trans} = \frac{2}{\hbar}\left\{\sum_f |\langle f_{exc};0_{exc}|\hat{V}_{int}|i_{exc};0_{exc}\rangle|^2 \delta(\hbar\omega_{exc}-\hbar\omega_f)\right\} \tag{5.27}$$

where $|i_{exc};0_{exc}\rangle$ is the initial state with an exciton in the donor and zero exciton in the acceptor; $|f_{exc};0_{exc}\rangle$ is the final state with an exciton in the acceptor and zero exciton in the donor; $\hat{V}_{int}$ is the exciton Coulomb interaction operator; and $\hbar\omega_{exc}$ is the exciton's energy. Neglecting the coherent coupling between excitons, i.e., the initial and final states can be written as $|i_{exc};0_{exc}\rangle = |i_{exc}\rangle|0_{exc}\rangle$ and $|f_{exc};0_{exc}\rangle = |f_{exc}\rangle|0_{exc}\rangle$, and the Fermi's Golden Rule can be approximated by

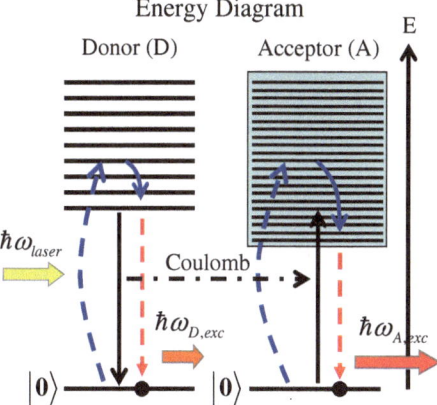

Fig. 5.7 Energy diagram for the directional process of exciton transfer from the donor to the acceptor. *Blue dash lines* represent the absorption process of the nanostructure (donor/acceptor). *Blue solid lines* denote fast relaxation process. *Red dash lines* illustrate light emission process (relaxation from the lowest excited state to ground state). *Black solid lines* represent the energy transfer from the donor to the acceptor. *Horizontal solid black line* illustrates the Coulomb interaction between the donor and the acceptor [reprinted (adapted) with permission from Ref. 9 (Copyright 2008 American Physical Society)]

$$\gamma_{trans} = \frac{2}{\hbar}\left\{\sum_f |\langle f_{exc}; 0_{exc}|\hat{V}_{int}|i_{exc}; 0_{exc}\rangle|^2 \delta(\hbar\omega_{exc} - \hbar\omega_f)\right\} \tag{5.28}$$

$$\gamma_{trans} \approx \frac{2}{\hbar}\left\{\sum_f |\langle f_{exc}|\hat{U}_{int}|0_{exc}\rangle|^2 \delta(\hbar\omega_{exc} - \hbar\omega_f)\right\} \tag{5.29}$$

where $\hat{U}_{int} = \langle 0_{exc}|\hat{V}_{int}|i_{exc}\rangle$ is the potential energy created by the exciton. With the help of the fluctuation dissipation theorem (FDT) [7] and the formalism given in elsewhere [8, 9], the Fermi's Golden Rule can be simplified into

$$\gamma_{trans} \approx \frac{2}{\hbar}\left\{\sum_f |\langle f_{exc}|\hat{U}_{int}|0_{exc}\rangle|^2 \delta(\hbar\omega_{exc} - \hbar\omega_f)\right\} \tag{5.30}$$

$$\gamma_{trans} = \frac{2}{\hbar}\left\{\frac{1}{2\pi}\int_{-\infty}^{\infty} \exp(i\omega_{exc}t)\langle 0_{exc}|\hat{U}_{int}(t)\hat{U}_{int}(0)|0_{exc}\rangle dt\right\} \tag{5.31}$$

$$\gamma_{trans} = -\frac{2\pi}{\hbar}\left\{\frac{1}{\pi}Im[F_{exc}(\omega_{exc})]\right\} \tag{5.32}$$

where $F_{exc}(\omega_{exc})$ is the response function given by

$$F_{exc}(\omega_{exc}) = \int dV \rho(\mathbf{r})\Phi_{int}(\mathbf{r}) \tag{5.33}$$

Here $\rho(\mathbf{r})$ is the local non-equilibrium charge density and $\Phi_{int}(\mathbf{r})$ the effective electric potential created by the exciton. Since the charge density is given by $\nabla \cdot (\varepsilon(\mathbf{r}, \omega)\mathbf{E}(\mathbf{r}, \omega)) = 4\pi\rho(\omega)$, the response function can be written as

$$F_{exc}(\omega_{exc}) = \int dV \rho(\mathbf{r})\Phi_{int}(\mathbf{r}) = -\int dV\left(\frac{\varepsilon_A(\omega)}{4\pi}\right)\mathbf{E}_{in}(\mathbf{r}) \cdot \mathbf{E}_{in}^*(\mathbf{r}) \tag{5.34}$$

where $\varepsilon_A(\omega)$ is the dielectric function of the acceptor and $\mathbf{E}_{in}(\mathbf{r})$ is the electric field inside the acceptor, induced by an exciton in the donor. Finally, the energy transfer rate from the donor to the acceptor is given by

$$\gamma_{trans} = \frac{2}{\hbar}Im\left[\int dV\left(\frac{\varepsilon_A(\omega)}{4\pi}\right)\mathbf{E}_{in}(\mathbf{r}) \cdot \mathbf{E}_{in}^*(\mathbf{r})\right] \tag{5.35}$$

where $\mathbf{E}_{in}(\mathbf{r})$ includes the effective electric field created by an exciton in the donor. Here, the electric field is calculated by

$$\mathbf{E}(\mathbf{r}) = -\nabla \Phi(\mathbf{r}) \tag{5.36}$$

The electric potential, $\Phi(\mathbf{r})$, which is needed to compute γ_{trans} (5.35), should be expressed as a total potential created by the electric potential of an exciton (on the donor side)

$$\Phi_\alpha(\mathbf{r}) = \left(\frac{ed_{exc}}{\varepsilon_{eff_D}}\right) \frac{(\mathbf{r} - \mathbf{r}_0) \cdot \hat{\alpha}}{|\mathbf{r} - \mathbf{r}_0|^3} \tag{5.37}$$

where ed_{exc} is the dipole moment of the exciton and ε_{eff_D} is the effective dielectric constant of the donor, which depends on the geometry and the exciton dipole direction, $\alpha = x, y, z$. Note that, to estimate the FRET rate, we need to calculate the effective electric potential due to an exciton in the donor in the vicinity of an acceptor.

The average FRET rate is calculated as

$$\gamma_{trans} = \frac{\gamma_{x,trans} + \gamma_{y,trans} + \gamma_{z,trans}}{3} \tag{5.38}$$

where $\gamma_{\alpha,trans}$ is the transfer rate for the α-exciton ($\alpha = x, y, z$).

References

1. P.L. Hernández-Martínez, A.O. Govorov, H.V. Demir, Generalized theory of Förster-type nonradiative energy transfer in nanostructures with mixed dimensionality. J. Phys. Chem. C **117**, 10203–10212 (2013)
2. C. Kittel, *Introduction to solid state physics*, 8th edn. (Wiley, New York, 2005)
3. P.Y. Yu, M. Cardona, *Fundamentals of semiconductors: physics and materials properties*, 3rd edn. (Springer, New York, 2005)
4. D.L. Dexter, R.S. Knox, *Excitons* (Interscience Publishers, Geneva, 1965)
5. L. Banyai, S.W. Koch, *Semiconductor quantum dots* (World Scientific Press, Singapore, 1993)
6. U. Woggon, *Optical properties of semiconductors quantum dots* (Springer, Germany, 1997)
7. P.M. Platzman, P.A. Wolf, *Waves and interactions in solid state plasma* (Academic Press, New York, 1973)
8. A.O. Govorov, J. Lee, N.A. Kotov, Theory of plasmon-enhanced Förster energy transfer in optically excited semiconductor and metal nanoparticles. Phys. Rev. B **76**, 125308/1–125308/16 (2007)
9. P.L. Hernández-Martínez, A.O. Govorov, Exciton energy transfer between nanoparticles and nanowires. Phys. Rev. B, **78**, 035314/1–035314/7 (2008)

Appendix

Förster Resonance Energy Transfer: Dipole-Dipole Mechanism

- $^1D^* + {}^1A \rightarrow {}^1D + {}^1A^*$: Singlet-Singlet Energy Transfer.
- $^1D^* + {}^3A^* \rightarrow {}^1D + {}^3A^{**}$: Higher Triplet Energy Transfer. This type of transfer requires overlap of the fluorescence spectrum of the donor and the T-T absorption spectrum of the acceptor. In this case, both donor and acceptor are in the excited states, but FRET formalism remains valid, with a few adaptations.
- $^3D^* + {}^1A \rightarrow {}^1D + {}^1A^*$: Triplet-Singlet Energy Transfer. This type of transfer leads to phosphorescence quenching of the donor.
- $^3D^* + {}^3A^* \rightarrow {}^1D + {}^1A^{**}$: Higher Triplet Energy Transfer. This type of transfer requires overlap of the phosphorescence spectrum of D* and the T-T absorption spectrum of A. The donor and acceptor are both in excited states.

Förster Resonance Energy Transfer Rate:

$$k_T(r) = \frac{1}{\tau_D}\left(\frac{R_0}{r}\right)^6 \tag{3.2}$$

$$R_0^6 = \left(\frac{9000(\ln 10)Q_D\kappa^2}{128\pi^5 N_A n^4}\right)\int_0^\infty F_D(\lambda)\varepsilon_A(\lambda)\lambda^4 d\lambda \tag{3.6}$$

Energy Efficiency:

$$\zeta = \frac{k_T(r)}{\tau_D^{-1} + k_T(r)} \tag{3.7}$$

$$\zeta = \frac{R_0^6}{R_0^6 + r^6} \tag{3.8}$$

© The Author(s) 2016
A. Govorov et al., *Understanding and Modeling Förster-type Resonance Energy Transfer (FRET)*, Nanoscience and Nanotechnology, DOI 10.1007/978-981-287-378-1

Förster-type Nonradiative Energy Transfer (FRET) Rate:

$$\gamma_{trans} = \frac{2}{\hbar}\operatorname{Im}\left[\int dV\left(\frac{\varepsilon_A(\omega)}{4\pi}\right)\mathbf{E}_{in}(\mathbf{r})\cdot\mathbf{E}_{in}^*(\mathbf{r})\right] \tag{5.35}$$

Dexter Energy Transfer: Exchange Mechanism

- $^1D^* + {}^1A \rightarrow {}^1D + {}^1A^*$: Singlet-Singlet Energy Transfer.
- $^3D^* + {}^1A \rightarrow {}^1D + {}^3A^*$: Triplet-Triplet Energy Transfer. This type of transfer is possible because the exchange mechanism does not imply transition moments of the donor and acceptor.
- $^3D^* + {}^3A^* \rightarrow {}^1D + {}^1A^*$: Triplet-Triplet Annihilation. This type of transfer part of the energy resulting from the annihilation allows one of the two partners to return to the singlet state from which fluorescence is emitted, but with a delay determined by the triplet state lifetime.

Dexter Energy Transfer Rate:

$$k_{Dexter} = \frac{2\pi}{h}KJ'\exp(-2r/L) \tag{3.9}$$

$$k_{Dexter} = k_0\exp\left[\frac{2(r-R_C)}{L}\right] \tag{3.10}$$